Quantum Consciousness
And Your Immortality

Quantum Consciousness And Your Immortality

By
James L. Forberg

Edited By
Richard A. Forberg Ph.D.

To order additional copies of this book, contact:
Xlibris Corporation
1-888-795-4274
www.Xlibris.com
Orders@Xlibris.com
35453

CONTENTS

PREFACE

T he intent of this book is to reduce to a practical amount the reading necessary for quantum awareness. At the same time we will venture into information about our physical and spiritual realities.

Modern quantum theory has removed and replaced Newton's view that the universe is a deterministic clockwork mechanism. It remains the function of our people of science to unravel the mystery of reality.

For all of my adult life I have been following the progress in the various fields of science by reading the journals, periodicals and books that are published in these fields. These highly trained ladies and gentlemen publish their work so that they can get some feedback from other individuals in their field. Some are easily readable by the general public because they leave out the very complicated mathematical computations that are necessary to arrive at their final result. Since I; and I suspect many of you find higher mathematics incomprehensible, we have to trust that these computations have been done correctly. You can rest assured that if there is an error that their fellow scientists would complain loudly.

With these facts in mind, the following pages will be a report of the progress science is making in the search for a location of our *consciousness*.

Sprinkled throughout will be a few of my own ideas, which I will indicate as mine.

Consciousness is words that can be replaced by others and still make perfect sense. Awareness is one as is soul and spirit. Most of the languages of the world have their own word or groups of words that signify the essential life force that we think of as *my* awareness of things around us.

Every intelligent individual at some point in their life will begin to wonder about the why, how and what of their existence. Why am I here? Why are things as they are? Why is it necessary to die? How can I avoid that fate? What is it like to die? What happens then?

Many of the religions of the world tell us they know the answers to these questions but offer no proof. With many of them even looking for proof constitutes a great evil.

The science of today is willing to search for the truth even though there are many different ways and competing ideas.

Quantum science and research, while far from ending the questions, are at least asking them. The results obtained from the various experiments give new directions in the search for truth. The quantum world of the electron, proton and neutron is still a mysterious place but is under close scrutiny.

Quantum mechanics and particle physics are increasingly finding evidence that there is a genuine connection between consciousness and matter. A measured electron can appear as a particle or as a wave but never both at the same time. Quantum particles seem to be mathematical points of possibilities. In other words, at that point there is no definite *physical* location or velocity for that particle. All things are possible. The human body is made of these quantum particles (or points). These particles occupy not only our four dimensional space but also the quantum dimensions of possibility and information.

All levels of consciousness shape our material world whether they are human, animal, plant or molecule. All of these things have the ability to be aware in varying degrees. This will be discussed on later pages of this book. By virtue of our quantum components we are residents of many dimensions. Our extension into space-time serves as a school of evolution. This means simply that choice and experience advance us physically and spiritually. Our bodies are temporary receivers and transmitters of information. Some received from the quantum realm and some sent from our physical space-time dimension.

The dissolution of our four dimensional body is a result of shifting quantum functions and is a temporary release to contemplate new forays into physical dimension. Physical life and death are the soul's classroom and graduation. It is the classroom of choice and experience. We attend many classrooms and have many graduations.

Although physicists are in agreement that our bodies are quantum entities, they disagree among themselves if we follow the rules of the quantum dimensions, or if those rules apply only at the sub atomic areas. Their reasons for questions are the mathematical difficulties with wave length, mass and etc.

In my studies I habitually visualize my body composed of trillions of quantum points. These points when viewed up close reveal quantum aspects. A photograph when viewed up close is also a group of points. As you back away from the points they tend to merge and gradually reveal something different. A photograph is a multitude of shaded points. A human body is a multitude of quantum points of possibility. Both views show that we must examine the small and the large to determine reality. A close view will reveal the components. A distant view will reveal the result of the combination.

An electron can be considered a quantum portal to dimensions beyond space-time. The addition of many more portals in various atomic arrangements does not change the electrons basic function. This function is as a point of contact between our physical space-time world and the higher dimensions of spirit and consciousness.

CLASSICAL AND QUANTUM QUESTIONS

This century has shown a gradual shift from the physics of Newton to the more esoteric physics of the quantum. For the most part, this change has affected only those areas of science dealing with sub-atomic particles.

It is necessary that we compare quantum physics with the classical Newtonian physics that most of us studied, or at least were exposed to in our school days. I will attempt to show that quantum particles rule our existence between the very large and the sub-atomic.

Newtonian physics is still capable of describing the world that we sense. This shouldn't come as a surprise at all since its rules have been obtained by employing our senses and brain. Newton used his eyes and brain when he saw an object fall. It may not have been an apple, but whatever it was started him thinking. He wondered why everything always falls to the earth. The outcome of his thoughts was our rules for gravitation. These rules have since been clarified and modified by Einstein and others, but the original thought remains. We have learned much using classical Newtonian physics. In our search for reasons for our existence we have discovered that our planet is between four and five billion years old.

Our planet, as well as our solar system, holds no special place in our galaxy or in our universe. In the universe there are probably billions of planets and galaxies that are very similar. If you can find a place on this planet away from the city lights, cast your eyes upward for a few moments on a clear night. If you have fairly good eyes you can probably see 2000 to 3000 stars (suns, with a few local planets) that are a part of our galaxy. That is just a very small part of approximately ten billion stars that make up our Milky Way galaxy.

In our space-time dimension we must always be aware of time and distance. Using the classical laws of Newton and others, we have been able to calculate times and distances between concentrations of matter in our solar system, galaxy and universe.

The most convenient way to measure space has turned out to be a yardstick of time. With diligent science the velocity of light was established at 186,000 miles per second, and used as a standard of distance.

Light will travel approximately six million million miles in one year, so that distance equals one light year. Keeping that figure in mind think about the fact that our galaxy is about 60,000 light years from edge to edge, and several light years thick and your imagination will be stretched to the limits.

If we could travel at light speed, it would take us about 4.27 light years to reach the stars nearest to us. These stars are Alpha and Proxima Centauri. Dr. Einstein's law of the interchange of mass and energy ($E=MC^2$), makes it *physically* impossible to travel at the speed of light. As we increase velocity, our apparent mass also increases. This increased mass requires even more energy to achieve a greater speed. Rather straightforward high school mathematics will show that it will take an infinite amount of energy to push matter to the speed of light.

There are many millions of galaxies like our own in the universe. Possibly even an *infinite* number of galaxies. These galaxies average 1,500,000 light years apart and all continue to separate (with the exception of the Andromeda Galaxy) at speeds varying proportionately to the distance from us. These facts have been gathered using a spectroscope in conjunction with a telescope to measure the shift in light spectrum from distant galaxies. This red shift or Doppler Effect is the result of the star or galaxy moving away from us.

These measurements have convinced cosmologists that the universe is expanding. If you mentally reverse time by calculating the trajectory of these galaxies in the opposite direction, it becomes obvious that the expansion is the result of a point beginning. This is the basis of the "Big Bang" theory.

This picture of a giant explosion from nothing isn't necessarily in conflict with the Christian bible but there are some time sequences that need some clearing up. The Bible seems to consider the age of the universe as only thousands of years. Science considers the age at 13 to 15 billion years.

Classical physics has given us a solid basis on which to evolve laws for physical systems (matter). Quantum physics on the other hand requires that we begin trying to understand non-physical occurrences and entities.

The physics of the past has been compiled by paying good solid attention to our sense impressions. The advent of quantum theory however has put a strain on the credulity of our senses. Much of this incredible change in the scientific view has bypassed the general public. Mainly this is due to the incredibly complex mathematics involved. Many scientists and science writers as well as other fields write only for the other members of their profession. This isn't necessarily out of secrecy or malice, but merely in an attempt to describe their projects fully. The answers that these professional men are getting from their tests and experiments are as confusing to physicists as they are to us. In the pages of this book I will report on the progress in quantum physics without the confusion of mathematics. As you read you will also get a background understanding of the question of human immortality.

We are aware from our classical physics that mechanical travel from one system to another is very unlikely. There is a possibility though that quantum physics can show us a way to bypass those great distances through control of our consciousness. Given the fact that all physical things (you and I included) are constructed of quantum points of possibility, and that these points aren't considered physical "things", it is also a fact that our natural state is non-material.

We live in our time-obstructed world and also in a timeless realm in the quantum dimension. Control of consciousness isn't as difficult as most people think and it can lead to great things for humankind. Later in this book I will show how to make changes in your level of consciousness.

The advent of quantum physics kind of knocked the legs out from under our platform of belief. Previous to sciences very close study of the atoms and their structure, we could pretty much rely on our senses and Newtonian laws. We felt secure that what we sensed was true from the microcosm to the macrocosm. It was sort of "what you see is what you get" situation. I personally am pleased that this is no longer so. Without our new knowledge of the strange things that happen at the atomic level, many questions about our consciousness would remain unanswered.

Nick Herbert in his book "Quantum Reality" enumerated eight variations (or interpretations) of reality as a consequence of quantum physics. These scientists are speaking of *experimental* results at the atomic level. As you will see these rules of reality *can* be extended to the macrocosm (our sensory world, or our level of space).

Niels Bohr's institute (a group of researchers) at Copenhagen tells us that there is *no deep reality*. In their opinion our senses are constituted to give us an impression of a material world, but that this reality is a reflection of something of a different nature.

This has been named the "Copenhagen Interpretation." It is the prevailing belief of establishment physics. This is an indication that many of the world's best minds agree that our reality is only a part of the truth.

Werner Heisenberg put his stamp of approval on the Copenhagen interpretation when he wrote. "The hope that new experiments will lead us back to objective events in time and space is about as well founded as the hope of discovering the end of the world in the unexplored regions of the Antarctic."

Part two of the Copenhagen interpretation states that we create reality by observation and that there is no *reality* without observation. It is but a short jump from that statement to one that says that the world (meaning universe) is but events of the mind. The question is whether that mind is mine, yours, ours or Gods. My personal preference is all of them for reasons that will become clear as the book progresses.

Part three states that the act of observation bonds the observer and the observed. This means that anything that you observe becomes part of your reality. Part three also states that all things are connected and act as a unit. This means that you and I and all things in our physical universe are one thing. This connection is undiminished by time or space. It also states that an action at on end of the universe is immediately available at the other. Our scientists can make a statement like that because the elements they are working with (quantum waves) are without mass and therefore without time or space.

Section four of the interpretation tells us that by experimentally observing the velocity of an electron, we find that the location becomes less and less clear. This experimental outcome has opened up the *many worlds'* interpretation of quantum reality. Some feel that in a situation where several different outcomes are possible that all of the outcomes occur and that an entire new universe comes into being for each outcome. (It is good to remember here that all of these universes, even ours are only *possibilities*.)

Many science fiction stories use this interpretation as a basis for their plot line. It can be overwhelming to think that there are an infinite number of yourselves doing different things at the same time.

The fifth interpretation says that it is pompous of us to believe that the universe should behave as we think it should. Quantum logic theorists believe the universe obeys non-human reasoning. They tell us that we should learn to think quantum logically. General Relativity changed the concept of space-time geometry. Quantum science demands that we stop using our old logic and concentrate on experimental results. Our common logic continues to lead us to expect Newtonian results. We should actually develop new quantum logic. This logic would have to be based on the premise that everything in the universe is one unit and that what we sense as physical is in reality constructed from non-material waves. Portions of these waves are made to seem material by one or more consciousness.

The sixth school of quantum reality does believe that the universe really is made of ordinary objects. (Meaning solid pieces of *stuff*) This neorealism feels that *things* have attributes of their own whether they are observed or not.

This contradicts such giants as Heisenberg, who said, "Atoms are not things" and Bohr who said, "There is only an abstract quantum description." Neorealists accuse the majority of quantum physicists of obscuring simplicity with mystification. The ranks of neorealists have within their group many well-known experimenters in physics. Einstein had a passion for realism and felt that sooner or later experimentation would bring quantum physics back to logical reality. He was joined by such names as Max Planck, Erwin Schrodinger and Louis de Broglie in this feeling.

Einstein never gave up his hope that somehow quantum physics could remove itself from the realm of probabilistic science and return to solidity.

The seventh interpretation of the quantum question is composed of a section of theorists and mathematicians that believe that only consciousness can create reality. This view brings with it the problem of establishing the parameters of *consciousness*. A prime consideration for consciousness is an ability to be aware of external forces, dangers and such. Cellular life can fit into that category. Even plant cells could pass the test. After all we do consider cells and plants *alive* don't we? It is true that their awareness is very rudimentary, but cell do sense (in a chemical way) light, heat, foreign cells, pH condition in liquids and many other states of matter that can be good or bad for their survival. Awareness (consciousness) can be reduced to very minimal states.

In the eighth possibility for the quantum basis, Werner Heisenberg divides the universe into real and semi-real. He considers the elementary particles to be only potentialities or possibilities. They are made *real* only by the act of observation.

Heisenberg's world of potentia contains all of the possibilities of any action. In the past following the rules of Newtonian physics, we like to think of a straightforward *action* and *reaction*. If we pay attention to quantum rules however any action has the possibility of a multitude of reactions. In the not real world of potentia, all of the reactions are present. Only one of these possibilities manifests in the real world. This possibility is made solid (physically apparent) only by consciousness. We must not forget at this point that any action also involves a very large number of conscious entities. We may sometimes believe that we make decisions and act on our own, but this is seldom true. The people, plants, animals and all things that are considered as part of the action, will have input into the degree of control of the action and reaction. This depends on each entities ability to phase quantum waves of possibility so that they add amplitudes enough to make them apparent in our physical world. (More on that later)

This multiple input into the cause and effect equation will be part of the discussion of our personal connection with more than one universe. The two that we are vitally interested in right now are the one that has many limits (our physical universe) and another in which all things are possible.

Sleep and Other States of Consciousness

Are we immortal? Is there more to us than a material entity? Is it possible for us to become more aware of the soul that inhabits the body? Other chapters will deal with some of the possibilities that have become more apparent with the advent of quantum physics. For now let's investigate something that has been used for centuries.

Altered states of consciousness such as Hypnosis, Meditation, Yoga, and biofeedback practices end in the same state of mental disassociation.

As a therapist I have used hypnosis for about thirty years and am constantly amazed at its usefulness as a tool of self-help and self-control. I use those terms purposely to indicate and emphasize that altered states aren't externally controlled but are an internal manipulation. All hypnosis is self-induced. A person is free to alter his/her state or not depending solely on their attitude at the time. A Hypnotist is really just an instructor that helps you to alter your state of consciousness.

There are still many people and many organized groups of people that are afraid of hypnosis or believe it to be evil or dangerous. Nothing could be farther from the truth. You may as well be afraid of day dreaming or playing the piano or preparing to go to sleep. These things all require various changes in your level of consciousness. Hypnosis is so natural that you do it many times a day just to accomplish the things that you must every day. These things include riding a bike, or playing an instrument or using math skills from your memory. Conscious life *requires* that we alter our state of awareness continuously.

The main question is why do we find it necessary to sleep? There doesn't seem to be a firm answer yet. We do know that there is a build up of certain chemicals in the blood stream that can be one cause of our drowsiness. Each of us can attest to the fact that after a certain period without sleep, every cell in our body cries out for the change to sleep. Medical science knows that sleep is necessary for our mental and physical health. Without it we become physically exhausted and mentally unbalanced.

Is it necessary for us to dream? Are these exercises of our consciousness to aid in problem solving? These dream events that may not have happened, or may not happen in the future could be our consciousness using *virtual* situations to prepare us to make the proper choices in our waking state. I chose the word "virtual" purposely because it connects directly with the spiritual part of our existence.

What really happens when we go to sleep? We are aware that our eyelids close and our glandular secretions fall to a low point. Respiration and heart rate slow down perceptively. Our brain waves (as indicated by an EEG) change character and we become unconscious. This state of unconsciousness is modified by the fact that certain stimuli can cause a return to consciousness quickly. This is not the case when drugs or injury are the cause of the unconscious state.

In sleep, the EEG will indicate many different wave shapes and amplitudes. Remember that these are all electrical (electron) functions, and are information coded in a binary system. One wave shape is so recognizable that it has been given a name. Our instrument at the back of the head and also at the forehead can intercept the Alpha wave most easily. This wave varies a little with individuals, but will average about nine to ten cycles per second. It has been learned through experimentation that this wave indicates relaxation and disassociation. This is true both physically and mentally. During this time there are no unusual sense impressions. The eyes are probably closed (not always) and the brain is functioning in an idle condition.

While the process of relaxation is underway in the physical body, there are complimentary changes occurring in that part of our consciousness that occupies the quantum dimension (our spirit). This seems inevitable because of the electro magnetic variations impressed on the physical particles as well as on the virtual particles that are in adjacent space and in between all of the material particles of the body. Keep in mind that the electrons that are racing around in our brain conveying information are also a part of those other dimensions.

In the conscious state the wave traces on the EEG are complex and change with sense input and motor output. When we relax and close our eyes as we do in preparation for sleep the complexity will decrease and the Alpha wave may appear in short bursts. Even with the eyes closed, if the brain is given a problem to solve the wave complexity will increase, and the Alpha will be minimal.

As we approach sleep the indicated brain waves become longer and slower. In deep sleep the waves will register from one to three cycles per second. The appearance alone would suggest that the slow cycle of sleep is a natural variation of energy level. As we monitor the waves of deep sleep we will notice intermittent bursts of activity that are called spindles. It is my opinion that this is a sampling system of the quantum consciousness. In this unconscious state the systems of the body are in their automatic mode, but are unattended by the conscious mind. An unexpected emergency of the body or its surrounding area would need the attention of the conscious brain activity. It is perhaps significant that the spindle can be most readily measured in the front part of the brain it is from this location that individual traits and emotions are controlled and the Astral body is said to emerge during astral projection. Since the spindle isn't triggered by any external sensation, the reason for its emergence must be a safety mechanism. During its short duration it will sample the electrical systems of the body to make sure that the organs are all functioning properly and that there are no sense systems warnings occurring. I feel that this spindle that occurs at regular intervals is the electrical indication of the quantum consciousness checking to make sure that the body is safe and functioning during the unconscious state. We do the same with satellites that we send to other worlds. Ground control will check the circuits of its charge at regular intervals.

What causes sleep? We all recognize for instance that monotony or disinterest bring about drowsiness and sleep unless one or more of our senses receive an impression that is stimulating to the brain and therefore to the body. By stimulating I mean an input that necessitates correlation of data in order to initiate the proper action.

I compare the brain to a transceiver in our satellites. The brain receives information about its surroundings in space. After comparing new information with old (memory), the brain sends it on via its quantum connection to the dimensions that await its information. A lack of stimulation will bring about that drowsy feeling that is called the hypnologic state. If continued long enough this lack of stimulation will allow the brain to move more deeply into the altered states of consciousness and ending in the unconscious state we call sleep.

Hypnosis is achieved by mental and physical relaxation. Purposeful physical relaxation and concentration on inner rather than outer impressions accomplish this. Since sleep is a very deep altered state of consciousness, the brain is paying more attention to memory and other quantum information.

Our dreams are a combination of remembered information and new circumstances that require our consciousness to make decisions. If you remember your dreams, you will recall that they require some kind of decision and action. These decisions can usually be correlated with situations that are unresolved in the waking state.

There is agreement among laboratories studding the phenomena that the reason we sleep remains unknown Medical researchers feel that there must be an as yet undiscovered chemical control in the body that serves as a triggering mechanism. Remember that *all* of the functions of the body are electrical in the end. Chemicals are electrical combinations.

The major portions of the body don't need sleep. Periods of rest they do need, but not sleep.

Experimentation has shown that the brain *does* need sleep. For all of these experiments I recommend that you read: "The Third World of the Mind! Sleep", by Julius Segal and Gay Gaer Luce.

Throughout this book we will discuss the idea that the brain is an instrument whose job is to maintain the physical portion of the link between the body and our eternal consciousness. Is the reason for the sleep cycle that the chemical functions of the brain becoming depleted? Are there other missions that the consciousness must perform? Is the physical brain allowed to idle while these missions are accomplished? Is the spindle that is observed during sleep an indication that the consciousness is monitoring the body periodically to assure its proper functioning and safety?

If we constantly monitor the EEG from the wakeful period to the slow wave of sleep, we will notice that the rapid constant waves of wakefulness will change gradually to irregular little waves.

These are indications of the drowsiness that I spoke of before. It is during this period that we experience rather strange phenomena. We are neither *awake* nor *asleep* in this stage of our mental functioning. Our consciousness is accepting information from both worlds of our being. We may *talk* during this period and some of our words may make sense and some not. Visions may appear that seem quite real. I recall once during a short nap, I opened my eyes and saw a big rat on the wall. It startled me and I threw a pillow at it. The pillow hit only a blank wall. The rat was of some other reality. This is the same thing that can occur in various stages of hypnosis. Our bodies may twitch and jerk in an effort to follow some physical thing that is happening in dreamland. This is an indication that we are so near to the conscious state that when we move or speak in response to a dream that it also registers in our conscious mind.

Sleep is a natural unconscious state. In this state we are more able to become aware of the infinite waves of possibility that surround us and are within us that are of the quantum dimension. Our brain, in order to remain active during our sleep stage uses events that haven't happened and probably never will happen.

Sleep is one state of our consciousness that we are aware of because there is a very definite physiological change that accompanies it. There are however many changes in our level of consciousness that are so much a part of our functioning that we tend to ignore that there has been any change at all. During a search of our memories for instance, one of the brains more noticeable electrical functions (the Alpha wave) becomes even more apparent during these times. To state in a simple way a very complicated physiological function, memory retrieval requires a lot of attention to inner sensitivities and not much too outer senses.

Watch a good musician some time as they play. Close scrutiny of their eyes, face and body will indicate that they aren't consciously manipulating their hands, but have altered their state of consciousness. They are playing automatically by ignoring most of their conscious senses. Also watch someone trying to remember something and you will notice a tendency to roll the eyes upward to the right or left. The direction depends on the type of information they are looking for. Tests have shown that when we do this there is an increase of that Alpha wave in the brain. More Alpha indicates the brain is paying less attention to external physical things and is studying memory and internal senses.

Altered states of consciousness are part of our physical body's mechanism of life. We can't even go to sleep at night without practicing self-hypnosis.

PHYSICAL CHANGES
IN RELAXATION

Let's consider for a moment what happens when you decide that it's time to go to sleep. After the normal things such as brushing your teeth and changing into your sleep attire, we lie down and relax. This purposeful relaxation starts certain things happening in your body. Some areas of the motor section of the brain begin to reduce the flow of electro-chemical energy to many of the large muscles of the body. The decrease of electrical flow is noticed (chemically) by the reflex system. As you probably know, one of the jobs of the reflex arc is, to protect the body rapidly when there is some physical danger. For instance, if you lay your hand on something that is warm but not searing, your brain will receive that information in the amount of time it takes the nervous system to transmit it to the proper area. Subsequently your reasoning brain will make a judgment whether to move the hand or not. If the item you touch is hot enough to destroy tissue, all of the heat sensors in your hand will fire at once and create a sizable flow of electrical current. Since the flow of electrical energy through your nervous system is relatively slow, before your brain got the message HOT and was able to alert the muscles to move the hand, tissue would be destroyed. Over the thousands of years of trial and error our bodies have developed a surprising bit of protection for this type of danger.

This large burst of energy that comes from the heat sensors breaks down resistance in the local reflex arc, and applies it directly to the muscles. This all happens before the brain knows the object is hot. It isn't widely known that this little reflex system helps us to go to sleep, as well as altering our state of consciousness for other purposes.

Purposeful relaxation decreases the current flow to the muscles. The reflex system recognizes this. By increasing the electro-chemical resistance in the afferent system, it begins to decrease the amount of information flowing from our senses to our brain and reticular formation.

What is the reason behind these phenomena? Common sense will tell us that if you are mentally alert and curious about your surroundings it's difficult to go to sleep. As your body lies or sits quietly there is less need for sensory inputs. You're not going to, bump your head or catch a ball or any, of the numerous things our senses enable us to do or prevent doing. The more you relax the less information your senses are relaying to the brainstem reticular formation. In this area is a sort of "Y" connection for our physical sense systems. The majority of our sense information enters at the bottom of the "Y". Our sense of smell is the only exception, probably because far back in our history we depended on our sense of smell even when we were asleep. The other sense signals split at the "Y" juncture with both sides receiving some of the signal. One side feeds into areas of the brain that could be considered part of our unconscious centers. The other leg of the "Y" spreads out into consciously controllable areas of the brain.

Our body is relaxing and the muscles are getting less and less information from the brain. This change is noticed by the reflex arc (electro-chemically) whose response is to increase the resistance to sense reception in the afferent system. Evolution has seen to it that when we need sleep, we can at least partially disassociate ourselves from the outside world.

If you ever feel the need to test what I have just said, have someone tape your eyes open sometime when you are really tired. Then allow yourself to lie down and relax. Although you may not notice, after a while you are no longer seeing. The signals are there, but your sight centers are ignoring them. Some people because of disease or injury can't close their eyes and they can go to sleep as fast as we can.

The division of sense information in the Reticular formation has a plus side as well as a negative side. While we are active and awake the conscious areas of the brain get a lot of information. At the same time areas of the brain that could be considered as unconscious centers are getting the same information.

If you are rendered unconscious chemically, due to injury, or asleep, the unconscious is still receiving a reduced amount of information from your senses. This is good for your protection, but not always.

Thirty years ago conversations in an operating room were proved to be very detrimental to the patient. Negative statements said between doctors and nurses were picked up unconsciously by patients, and delayed or negated their recovery. Recently more care is taken in the operating room when discussing things they see or become aware of.

Some years ago there was a case that stressed this point very clearly. A prominent surgeon (who will be nameless), became ill himself, and sought the help of his fellow physicians. His medical knowledge indicated that the symptoms could mean cancer. He was taken to the operating room and opened up for an inspection. The surgeons found a tumor and in their conversation they said it looked suspicious. They removed the tumor and had it sent to the lab for a check. In the meantime, they closed up the abdomen and sent the patient to recovery. Subsequently the lab found that the tumor was not cancerous.

Some time later in the recovery room it was discovered that both kidneys of the patient had shut down and were not doing their job. The attending physicians couldn't understand why the patient wasn't regaining consciousness. One of the doctors, was also a practicing hypnotist, was suspicious about the cause of the kidney malfunction. Using hypnosis techniques the doctor finally got the patient to talk even though he was still semi conscious. He asked the patient why his kidneys were not functioning. The patient answered that death from ureic poisoning is much gentler than the slow painful death of cancer. When he heard this the physician informed the patient that the test showed no cancer in the tumor. Within minutes of this conversation both kidneys began functioning again.

Misinformation received in an unconscious state and a previously held opinion almost led to an unnecessary death. Many other problems can arise from this interconnection of the conscious and the unconscious. Activities of the unconscious can be impressed on the conscious brain. The manifestations of this condition can come in the form of hallucinations, mental voices and physical illnesses, all of these due to the break down of circuitry somewhere.

The illnesses that are classified as psychosomatic are in large part due to this crossover of information.

You should be aware now that you can alter your level of consciousness by relaxing. All hypnotists ask you to relax in one way or another.

This business of watching a pendulum or staring at something is merely a ploy to get you to tire your eyes so much that you have to strain a bit to keep them open. The hypnotist will notice this and say," you're becoming very tired and wish to close your eyes." It's a bit of psychological humbug that we should grow out of by now. We ask you to close your eyes only because it is easier for people who haven't practiced meditation, self hypnosis or biofeedback to alter their state with their eyes closed.

The systems used to alter your state of consciousness may achieve deep changes and some shallow but the various levels of consciousness can all be used for various therapies. I'm sure you will hear some practitioners say that their technique is better for one reason or another, but that is just business talking. The effectiveness of the abilities manifested in the altered state are a product of will and attitude on the part of the patient.

It should be apparent now that hypnosis isn't anything mysterious or dangerous. The process of inducing the altered state is totally within the control of the subject.

Self hypnosis may be quickly learned and is a great advantage to individuals who wish to make changes in habits and traits that have become automatic and are detrimental to health and happiness.

If you wish to alter your state and make some of these changes, simply find a comfortable quiet place where you will not be disturbed and sit, recline or lie down. Get in a good comfortable position and relax. This relaxation can be more easily achieved by using a soothing background sound such as recordings of wind in the trees, or morning birds, or maybe the sound of surf on the beach. If you don't have access to these recorded sounds just take any radio and search through the dial until you find a place that has no station. Generally these areas on the radio dial have static, or a hiss. This white noise will aid in altering your state of consciousness.

Many people that are untrained in altered states make the mistake of expecting a profound change in these transitions between states of consciousness. As you relax, do not expect to become unconscious, or unaware of your surroundings. This only comes with deep changes such as sleep, or the coma state. These are both altered states of consciousness, but more profound than you require for the process of self instruction.

As you sit or lie relaxed, begin to channel your thought processes by visualizing pleasant things. These mental pictures may be of a walk through the woods with the birds singing in the trees. If you prefer, you could take a mental walk on a beautiful beach with the sound of the surf in the background. This mental visualizing along with the muscle relaxation will begin to alter your awareness of the external world, and begin to enhance your awareness of your internal non-material world. You will reach a point in your hypnosis induction that indicates a disassociation with your body. Intellectually you will know that you are lying, or sitting somewhere but you really don't know the location or position of your arms or legs.

This is a good time to give yourself some instructions about changing those habits. You will simply indicate what you wish to change, and how you plan to make this change. In this altered state, your brain functions differently. You begin to have more control of those automatic habits, and to introduce changes that modify those actions.

Remember that habits are the result of repetitive mental and physical actions. The same is true for your attempt to change those habits. Whenever you have the time during the day or evening, repeat what you did before. Alter your state by relaxation and visualizing and repeat the instructions you gave yourself before. These repetitions will begin to make changes in those habitual activities that you wanted. It is just the same as learning to ride a bike or play the piano. Practice makes perfect and you will have established a new habit that you feel is right. Always remember that hypnosis isn't a weird feeling, it's a relaxed feeling. You may have witnessed someone acting strangely during a hypnosis demonstration, but that is show business. Those individuals knew exactly what they were doing at all times. The volunteers were just having fun with the audience.

Self control requires that we program our reaction to physical and non physical information that we receive in our brain. The best way to do this is to take control of some of those areas of the brain that are normally considered unconscious. By making these changes, you can begin to control such things as heart rate, blood pressure and sensitivity to pain. There may be no limit to the things we can control with our consciousness once we learn to do it. The connection of the consciousness and the quantum dimensions will be outlined and explained in the following pages.

DREAM TIME

In adults, experimenters find there seems to be a natural sleep cycle whose periods last roughly one and one half hours. These cycles vary widely, but this is a good average.

After a sleeper enters that big slow wave sleep that I have mentioned, he/she seems to remain there for that period and after a time though the EEG begins to change.

During a period of approximately ten minutes, a sleeper will move up from stage four delta sleep through lighter and lighter stages. Eventually the sleeper will reach the light sleep of stage one. Most of the time this stage one sleep will again be modified into an even lighter stage called *rapid eye movement*, or REM sleep.

This is the stage of sleep when you have most of the dreams that you remember. It is actually a hypnotic state during which you are very close to wakefulness. Usually if you don't begin to dream during this stage, you will wake up and wonder why you woke. I have shown many mothers and fathers how to take advantage of this state to give their young children suggestions that will modify bad habits and traits.

REM sleep is called that because during this stage the eyes move beneath the lids as if they are following movement. If sleepers are awakened during this phase and are questioned, they will invariably say they have been dreaming. Dreams that occur during this stage of sleep seem to be the most vivid or at least the easiest to remember. There is evidence shown by the EEG that dreams do occur during deeper stages of sleep but they are harder to remember. It seems that in those deeper stages that we are farther away from physical reality. Is it necessary for us to have more than one type of sleep? The sleep laboratories aren't ready to tell us yet, so we will turn to psychology and also the psychic sciences for possible answers.

Sleep scientists whose ranks include psychologists as well as neurologists have discovered that if sleepers are monitored and denied this REM sleep for a period of time that certain irregularities begin to occur. If this REM sleep denial is extended, these abnormalities could result in permanent damage. No one is certain if it is the dreams that are necessary, or just that portion of sleep in which dreams occur more vividly. Is it possible that during this period our consciousness is sort of clearing the circuits in preparation for another day? Is it also possible that during this period of sleep we are experiencing memories pulled at random from our unconscious? To me, neither of these things seems to be a good answer. The dreams that I have and those reported to me in sleep laboratories bear little resemblance to memories of actual experiences. They do seem to be distortions of concerns that are present during our waking life. Some of them (at least in my case) involve communication with individuals that are no longer alive. For me dreamtime is very pleasurable. There are times in my dreams when I can will myself to lift up above other people in the dream and float there feeling rather smug that I can do so. Psychology would say that this is an attempt for the dreamer to elevate his standing among his peer group. Possibly, but it sure is fun.

In our dream state we can be aware of physical things, non-physical things, or both at the same time. I have had many occasions in which physical sounds are incorporated into my dreams. Sirens, car horns and barking dogs seem to alter my dreams very rapidly to weave these sounds into the vision. Consciousness is awareness, but the unconscious state is still an area of awareness. At least in this REM stage.

I have said to my students many times that not only do a lot of dreams make no sense at all, but words and even sentences spoken during dreams also make very little sense.

Since I have a continuing interest in dreams and consciousness, some time ago I wrote down an example. My alarm clock went off one morning and apparently roused me from a REM period of sleep. A dream and a portion of conversation were very fresh in my mind, so I wrote it down for later study. What I wrote was "gee Jeffrey did a boob suspicious thing." The only thing that makes any sense to me at all is the name "Jeffery." I have a son by that name. As far as the rest of it, your guess is as good as mine.

Some (dreamers?) report out of body experiences during their sleep periods. Everyone who studies this phenomenon agree that an altered state of consciousness is necessary to achieve the separation from the confines of the body. It is highly possible that some sleepers use the various stages of sleep to accomplish this. These individuals always report that their floating body is always accompanied with a thin silver cord that stretches from their material body to their non-material body. It is my view that this cord is a visual representation of that spindle (spindle waves are propagating synchronized oscillations that are recorded during some stages of sleep). It indicates that the spiritual body is still in contact with and is responsible for the material body. While the spiritual (quantum) body is off on some adventure, it still must be able to check on the condition of the physical. Some travelers tell us that they become aware that if this silver cord were to become detached that the physical body would die. The spindle waves are very complex when they occur, which is an indication that there is a lot of information passing both ways through that silver cord.

To those of you who have had the occasion to observe a sleepwalker it will come as no surprise that they aren't *with us* during those wanderings. Their mental state is akin to a medium stage of hypnosis, where some control of the body remains. The mind however is creating a world of its own. It is significant that sleepwalkers as a rule do not remember their escapades. This is not always the case, but most do not remember walking or talking. If we equate somnambulism with hypnosis, it is well to remember that only in the deepest states of hypnosis is an automatic memory loss observed. Not surprisingly somnambulism is derived directly from deep Delta or stage four sleep and then they move up to a somewhat lighter stage to allow their muscles to function.

During a twenty four hour period, each individual will go through many different stages of altered consciousness. Memory retrieval, day dreaming, nighttime dreaming and problem solving are all various altered states. All of the various psychic sensitives use these various states to acquire information from the quantum realm of information.

STAR STUFF

Hopefully you are beginning to recognize that you possess a great deal more control of your body and consciousness than is generally accepted.

I want you to recognize also that this control is initiated by thought that specifically is the electro-chemical processes of the brain and its extensions. Since the chemical part of this process is really an exchange of energy in the form of ions, we can simplify our discussion by saying that the whole process is electrical. All material things are electrical in nature. Atoms are composed of negative, positive and in most cases neutral points of energy. The term *points of energy* must be defined a little more closely. In quantum physics an electron is a point particle because it occupies no space. This means that the electron is an imaginary mathematical point. There is a point of influence that does not occupy space in our universe. The same is true for the other quantum particles such as the proton, neutron and photon. If you happen to be looking for a specific location of an electron (for instance) you can narrow it down to a range of possibility with the proper equipment. When you do that however other things about the electron such as its velocity become indefinite. Recognize that our bodies are composed of various combinations of particles that by themselves are not considered as part of our space. Quantum science really does enter into our lives very intimately.

We are after all a product of the stars. When I was still lecturing and teaching I used to make it a point to explain this to my students. It is at the same time humbling and exalting to recognize that the majority of our body was actually manufactured in the Big Bang. The largest part of our body is water. A molecule of water is two parts of hydrogen and one part of oxygen. High energy physicists tell us that all of the hydrogen in the universe was produced a measurable time after the big bang. At that time you would find only Hydrogen (75%), Helium (approx. 25%) and traces of Deuterium. Consequently a large part of your body employs particles that have been around since the beginning of time and the physical universe.

The rest of the elements of your body were produced at the center of a huge sun, or suns. All of these heavier elements were spewed out into space in some explosions that are second only to the Big Bang. These explosions occur at the end of the life cycle of stars that are many times the mass of our sun. These huge balls of hydrogen have been so heated by their contraction due to gravity that their atoms begin to fuse. This fusion process forms new atoms of helium, and releases a tremendous amount of energy. The heat energy at the heart of the star is enough to counteract the force of gravity that is constantly trying to collapse the star. This fusion of the hydrogen atoms will continue for billions of years (depending on the mass of the star). There comes a time however when most of the hydrogen of the star has been fused into helium. At that time, the fusion begins to slow and the internal heat pressure begins to decline. The star once again begins to shrink because gravity has the superior force. As collapse continues the internal heat starts to rise once again. As the heat at the center reaches a critical point, the violence of the temperature begins to fuse the atoms of helium that were manufactured by the fusion of hydrogen atoms.

The star once again becomes stable and will radiate its excess energy into space.

For very large stars this process continues in intervals that are dependent on the mass of the star. The helium fuel will be exhausted as it is fused into carbon. Eventually gravity will again become more powerful than the internal heat and will begin to squeeze the star. This pressure will once again raise the internal temperature of the star. As if following a directive, the internal heat begins fusing the carbon into still another element higher on the periodic scale.

For the largest stars this battle between the force of gravity pushing in and the heat pushing out continues. When the fusion reaction reaches a point where the resulting output is iron, the star is near its end. The fusion process does not allow the iron atom to be transformed into yet another more massive atom. Consequently the core begins to lose heat as less and less atoms fuse. This loss of internal heat pressure will, at a time ordained by the mass of the star allows a sudden catastrophic collapse inward.

This inward rush of material from the outer levels of the star suddenly meets the dense iron core. The iron resists compression and as a consequence the implosion reverses itself and becomes a massive explosion called a Super Nova. This rebound is accompanied by compression temperatures that are sufficiently high to fuse some of the stars matter into elements even higher on our periodic scale of elements.

The Super Nova explosion blasts into space the materials necessary to build new stars, planets and you and me. The life cycle of stars is dictated by the laws of physics, but the material that these very large stars manufacture is needed to construct our world. The universe does seem planned doesn't it?

Material cycles occur for all matter. We have discussed that the matter of your body was fashioned either in the Big Bang (Hydrogen), or in the center of a massive star, and expelled in a Super Nova explosion.

Those atoms gradually drifted together to form the Earth that was orbiting a new star, our sun.

The atoms of your body have been many things before. Hydrogen and Oxygen combined to form the primeval seas and the more massive atoms combined in various ways to form the land.

One theory for the origin of life on our planet is that the warm sea lapping on the shore formed a suspension of just the right chemicals to form a brew. This brew of chemicals then borrowed some energy from the sun or possibly from the lightning to fashion combinations of elements. With time this elemental base began to aid in the construction of new material just like itself from the surrounding sea. Some of our atoms may have been involved in that. Once again, through the largess of time these self replicating groups of molecules became more and more complicated. By far the best addition they made to their bag of tricks was the ability to move about in their environment. This ability afforded them more control over their food supply and physical conditions. More time passes and small important changes occur. Many of the entities of the sea derive their energy directly from the sun through chemical conversion. These green things of the sea also have companions whose atomic structure is arranged differently. Instead of getting their energy directly from the sun, these creatures engulf and digest the small bits of life that *did* get their energy from the sun.

This arrangement of life still exists. It's called the food chain. From single cells to humans, most get their energy from the photons of the sun. Some deep sea creatures are an exception. The green plants get it directly, and the rest of us through various intermediate steps. Photons you remember are quantum particles that occupy no space. Like electrons, protons and neutrons they are mathematical points of possibility. Some of your atoms may have been involved in these tiny physical incarnations. Maybe one of your atoms contributed as part of the algae or the sea water. Perhaps an amphibian ate some of the algae and some of your atoms became part of its tissue. Sometime later visualize a larger animal devouring the amphibian. Once again there has been a transfer of vehicles for a tiny atom or two of the material that makes your present body. Remember that atoms themselves are almost indestructible. It takes other particles of great velocity or energy to smash an atom.

Our present bodies are composites of atoms that have been many things before. If for some reason you are looking for resurrection in the future to this same body, as the Christian bible intimates, there really are many forms that will have to re-exist again before that can happen. Some of these materials come from the blast of the big bang and some from super novas. All however finally come from the earth and the sea.

All existing matter whether rocks or man came from previously existing materials. Men and women create a child from a sperm and an ovum. The growing fetus manufactures its own tissues from the diet materials supplied by the mother. The form this body takes is governed by a tiny chain of chemical molecules that are arranged in an electrical code (DNA) each element sharing electrons to achieve the element necessary to perform its function.

The foods the mother eats that are transformed into the tissues of the fetus have come from the soil. Photons of the sun and the water of the planet supply the energy and fluid.

In a similar way when a person dies, he or she ceases to exist as a physical individual. The materials of the body do not simply lapse into nothingness. The chemicals of the now lifeless body return to the earth and atmosphere to be used again by other entities or to replenish the soil. This return is sometimes slowed by protective barriers around the body but eventually nature wins and reclaims its material. These chemicals (electrical arrangements) will be used in the future by other forms, or simply as a part of the structure of our earth. It is scientifically easy to state and prove that the building blocks of your body have existed for billions of years and some even since the beginning of time as indicated by the Big Bang.

Always remember that the matter of the universe is composed of electrons, protons and neutrons. These so called particles are also classified as quantum waves.

The elites of physics experimenters have just as much trouble as you and me picturing something that can be a particle and a wave at the same time. Physicists now like to say that a quantum is either a particle or a wave depending on what you measure for. The quantum will anticipate what you wish to know, and will show you that particular facet of its existence.

QUANTUM REALITIES

It is not easy to prove that the force that drives your body and contains your consciousness *has been* and always *will be*.

In the scientific community investigation into the forces of spirit or non-physical phenomena is politely referred to as metaphysics. In some circles the references are not nearly as polite. In Greek philosophy metaphysics means *after physics*, however it soon came to mean *those topics beyond physics*. Scientists, being the cautious people they are, always managed to say that investigation in these areas still may have a bearing on the nature of scientific inquiry. As of now metaphysics is really a study about the total field of physics. The question of reality has over the past half century become a real irritant to staunch materialists.

Quantum mechanics, a science that is new to this century, has cast doubt on our ability to fathom what is real and what is arbitrary. Werner Heisenbergs' "uncertainty principal" states that all measurable quantities are subject to unpredictable fluctuations and therefore an uncertainty in their values.

To test these uncertainties, observables are grouped into pairs such as "position and momentum" and "energy and time." It has been found that careful testing to ascertain the position of an electron for instance, will increase the uncertainty of its momentum. The reverse is true if you design experiment to find the momentum. In that case the position becomes harder to define.

If you want to be able to predict the future state of a particle (we are all made of them), it is essential that you know its position and momentum precisely. Since this experimental uncertainty seems to be always present, it appears that it is impossible to predict the future from the past. It is probably true however that in our macro world the immense number of particles associated in any activity will create a medium effect that is predictable. While this uncertainty always seems to be true experimentally at the quantum level, it is still possible on our scale (macroscopic) to have deterministic laws.

Some years ago during my lectures I would ask the students if they thought the past still exists: After some discussion they would generally agree that the past does still exist in the various kinds of radiation that were produced at those instants. These radiations would include heat, light and vibrations produced in the atmosphere and so forth. The discussions would become terribly complicated since all material things, even rocks produce vibrations specific to the elements contained in their structure. Our bodies do the same. We radiate in a wide range of frequencies.

We decided that the past really does exist in some residual fashion. The big bang exists in that same kind of way with the background radiation.

We would turn our attention to the present. Once again after a discussion we would get an agreement that the present is almost infinitely small. As you speak, the words are already in the past. The dividing line between past and future is extremely small.

Eventually we would get around to questions about the future. Does the future exist or is there nothing after the present? We are sure the past at least did exist since we have all of the artifacts. We have pictures, monuments, books, grandma and grandpa. Even if we discount our arguments about the radiations of the past, we are sure there was one. We are living the present and are aware of ourselves, other individuals and material things. All of these people and *things* around us help to define who or what we are.

If the past and future meet at an infinitely small juncture (present) is the future already there? Are we (as Shakespeare said) merely actors on the stage of life? Most western cultures don't like to think that way because that takes away our freedom of choice. If the future doesn't simultaneously exist, that would mean that everything, *the entire universe,* has to be recreated every instant. That doesn't make much sense when you consider the energy expended in those recreations.

Usually after some pushing and prodding my class participants would agree that perhaps there is a kind of compromise between the past and the future. Maybe the very large things have a semi-permanent existence. This would include the galaxies, stars, planets and mountains, not really a permanent existence, but one that is very slow to change. As you regress down the scale of size, change becomes more rapid and seemingly uncontrolled. Once again when you reach the areas of the atoms, photons and such, changes are so rapid that they register only as an infinity of possibility. Possibly that is merely the view we are allowed because of our size and sensory apparatus? If we were galaxy sized wouldn't human activity seem terribly fast and maybe pointless?

I always made a point to emphasize that all of the combined forces of the universe create the future. Your consciousness and mine are part of this array of force. The awareness of humanity and all living things together with the electrical, chemical and gravitational forces precipitate choices that mold the future from the quantum dimensions of possibility. Natures' forces such as the wind, the rivers and the sea have a great say in what the future will be.

Massive things for the most part make changes slowly with the exception of violent changes like supernovas.

Our place in the universe is just about half way between the size of a proton and the size of a planet. In our sphere we help to shape the future that we become aware of. When I say *we* I don't just mean *humankind*, but all things. As I have mentioned before all matter disturbs the universe in its own way. As humans we make decisions and choices, some of which are major but most are very minor and seemingly necessary to continue our day. Each move we make though begins to shape tomorrow. Each minute your spouse, friends, dog, your senator or president all are making choices. What they say or do will either add or subtract from what you wish for the future.

Things that we view as inanimate can modify the future. Suppose a bad winter has opened a pothole in the road that you travel. As you drive down the road, you hit the pothole with your tire. Maybe the tire will blow and you will have a repair bill as well as a bad temper.

All of the forces and pressures at least of our world push and pull on the clay of the future with the unpredictability of the electron. It is clear though that if we apply our intellect in a controlled fashion we can perhaps begin to mold that future clay more the way we want it.

For thousands of years much of humankind has felt helpless at the hand of gods or God. Since it is obvious that our hands and minds have a big part in shaping the future, does that mean that we are partners with God or Perhaps even part of the mind of God?

Paul Davies in, "The Mind Of God." writes:

"Why should human beings have the ability to discover and understand the principles on which the universe runs? In recent years more and more scientists and philosophers have begun to study this puzzle. Is our success in explaining the world using science and mathematics just a lucky fluke, or is it inevitable that biological organisms that have emerged from the cosmic order should reflect that order in their cognitive capabilities?"

It seems to be natural in our western culture that we tend to reject the possibility that God has written the future and that we are just following the script, like puppets on a string. The alternative is that the future is *uncertain* and becomes *real* only when all of the forces and pressures make it so.

Consciousness is awareness. In the brain awareness becomes more acute with change. Our brain functions best when there are changes. You may notice that if you stare at something for a while and there is no movement, your brain becomes uninterested and turns inward. At the same time the rods and cones of the eye begin to fire less and less in response to the lack of change of focus. This phenomenon is used as one of the basic meditation or self hypnosis techniques.

The concept of evolution has made us aware of many types of changes as a protective device. If we remember the position of a tiger in the past, and become aware that he is nearer in the present, we need to make changes in the future. Similarly if our spouse was happy yesterday and is not happy today, we need to search for the reason. All of these thought processes require the transfer of energies throughout our brain. This movement of quantum particles (waves of possibility) has an effect on us *now*, in the *future* and to everything else in the universe to varying degrees.

THE SPACE OF
CONSCIOUSNESS

T ry to think of this material as an extension of your understanding. The information gained here will in time alter your view of the physical aspects of life, and will begin to relax your tensions.

While we develop a better understanding of the universe as it really is, we increase our level of consciousness. Our thoughts are the elements that are used to construct the universe. All of the various levels of consciousness here on our world and elsewhere have built and sustain what we sense.

It is useful to use our ability to visualize from time to time. Not only in deep altered states but also during periods of recreational reading. Mysteries, romance novels, science fiction and westerns all take advantage of this ability.

There isn't anything mysterious about visualizing. We do it all the time. If you want to redecorate your living room, you arrange things in your mind first. If you want to sculpt an animal out of clay you will see what you want first in your mind.

The same is true of most of our actions. We follow the directives of our consciousness. Many times these directives change with circumstance but nevertheless we are activated by mental commands.

As our physical scientists continue to experiment and understand the nature of space-time, the need to use our ability to visualize becomes very important.

The big laboratories such as MIT, Bell, Princeton and many others with their sophisticated computers can use their machines as an aid in visualization. You and I have in our possession the best computer known. If we use our brain and mind properly, it will help us to more fully understand the universe around us.

Many times on these pages I will remind you that quantum particles occupy no space. To physicists they are not *things*. Try to visualize an electron all by itself. You will see in your mind a little point. So far that is okay as long as you remember that it is a mathematical point that occupies *no space*.

So far then we have this one little point all by itself with no space involved. You can't say that it is this far away from that, because there isn't anything else available.

Let us then envision one more electron some distance from the original. You can say that now because we have added one more point we have created the dimension of length. There can be an imaginary line between the two mathematical points.

As we get bolder we can add two more electrons to our field. We will place them so that when we visualize lines between all four points, the angles will all be right angles.

You can recognize now that we have created something with two dimensions (length and width). We could visualize a sheet of paper with no thickness.

Since we have this two dimensional something, let's have more fun by adding four more points to our field. Once again place these quantum points so that when you add imaginary lines from the new four to the old four they will all connect with right angles.

With this mental maneuver you have created a three dimensional space within the confines of the quantum points. This space consists of length, width and depth. You have made this space with your consciousness. It exists in your mind and brain.

Be aware that this imaginary space has its counterpart in our universe. In our so called physical space we measure from point to point whether it is in inches or light years. The points we measure from are quantum particles (points). These points are not within the space that is created. That space is only described by the quantum points. According to physicists, quantum particles occupy no space. They are instead areas of connection to other dimensions. What that means is that where ever there is a quantum particle (electron, proton, neutron and photon), that these points abut with and are a part of other dimensions.

Keep in mind that space can be any size. It can be the size of our universe, or the size of a hydrogen atom. The hydrogen atom has two quantum points. These are the electron and proton. Between these two points can be drawn an imaginary line that is the boundary of a dimension of space.

Our sense systems tell us that our body is a solid structure. The majority of our body really is what we consider empty space. The spaces between the particles and the particles themselves describe the size and shape of our space (body). You can visualize a three dimensional connect the dots kind of picture. It is Possible that when you contemplate the world around you, you think of yourself as *in here*, (your vantage point) and the rest of the world as *out there*.

Your consciousness builds for you a universe within a universe. This is your body that you use to traverse the dimensions of space-time. The size and shape of your body is described by the arrangement of quantum points in the elements of your tissues. These in their turn have been regulated by the DNA that we talked about before. Keep in mind as you read from this point on that each material thing can be considered a universe of its own. Its outline is a particular arrangement of quantum points. These points are actually areas of force that emanate from other dimensions.

It is my belief that the world of matter and of humankind is an extension of possibility from those dimensions that are the immortal home of our individual and collective consciousness.

Universal Spiritual Entities

To understand how we can be universal entities, it is necessary to think some about Quantum scientist John Bell's experiment that shows all quantum material to be one unit, universe wide.

An experiment done by Einstein, Podolsky and Rosen, labeled EPR originally was done using momentum correlated electrons that were diverging. (In plain language, two electrons moving at the same rate in different directions.) Since then however an easier way has been devised using photons that are polarization correlated.

With both of these methods of experimentation (electrons or photons), the answers are always the same. With paired electrons as well as paired photons if they are separated to any distance, they still react as one unit. Bell explains that there isn't just a waveform that connects the remote particles, but that a bit of each particles *being* stays with the other. Part of its phase remains in harmony with the other so that they are always in immediate contact. Space separation isn't important. It can be inches or light years, but the contact is still instantaneous.

You may recognize that this phenomenon is exactly what happens in a holographic system. In holography, each bit of a photograph (for instance) carries the major part of the whole. This phenomenon (holography) seems to connect each atom and even each electron as though they were one unit even though they are millions of miles apart.

Since all mater is ultimately a quantum system, there is immediate contact with phase connected entities. This can be used to explain a great many psychic happenings. In psychometry, for instance, if a sensitive individual handles a piece of material that someone else has had in his/her possession for a period of time, they can immediately be aware of the location and condition of that individual. The exactness of the information depends on the sensitivity of the psychic. A case of bi-location can be explained for the same reason. A very strong consciousness that is maintaining a physical body at one location can be responsible for another with the same phase at another location. One would not be aware of the other. The phenomena of a ghost could be explained because of this connection. In the quantum world of information, time is not a factor. If a person who is no longer living a material life left a very strong memory in the quantum information dimension, that memory could possibly still influence particles in our world to assemble a visible but not material body. Precognition in my view is still dependant on a trend rather than definite information. If a sensitive can move through the information of the quantum world and collect information that shows a trend towards a certain outcome, they can report that as a view of the future. Since we are aware that the future is dependant on decisions and actions, these views can be accurate, far from accurate, or somewhere in between.

Since all matter is ultimately a quantum system, there is immediate contact with all phase-connected entities. Basically that means *everybody and everything.* I include the animals and plants on our planet as well as else ware. Many experiments have been done with plants in which negative thoughts were directed at plants. Negative thoughts stunted their growth, but positive improved their growth. This information is in our scientific records. The same is true of animals. If we don't like them or have negative thoughts about them they are wary of us. On the other hand positive, loving thoughts bring friendly overtures from them. Much has been written about positive results of prayer. A prayer is a thought process in which you ask for help for yourself or others from a greater power. It follows that if any number of individuals pray for the same thing (follow the same line of thought) the amplitude of this prayer becomes greater in the quantum dimension. (I will repeatedly refer to this as the spiritual dimension.) The result of this is a greater probability that this thought will be accomplished. The prayers can be said at many places on our globe, but the effect is the same.

Non-local action such as this has in the past been dismissed as impossible. Einsteins General Theory says specifically that faster than light communication is impossible. His equation is based on conversions of mass and energy. If a condition existed in which transport was external to the rules of space time, contact could be instantaneous.

Using the same theoretical process, we could bypass the rules of space-time by consciously disassembling our bodies and re-assembling them elsewhere. It will take a much better control of our consciousness than most of us possess now.

If we pay attention only to the old rules of physics, quantum waves (and particles) emanate from nowhere. We are now aware that electrons, protons, neutrons and photons do not originate in our physical universe. Since their environment is outside of our universe, time and space are meaningless to them as individual particles. A combination of more than one particle would constitute an element and confine them to space time. As Paul Davies says in "old and the new physics" "science which by definition deals only with the physical universe, might successfully explain one thing in terms of another, and so on but the totality of physical things demands an explanation from without."

Quantum physicists consider particles as *information* that is brought to definition by the combined consciousness of all of the entities that are aware of these points of radiated quantum fields. This means that humankind, the animals, plants, cells, viruses and anything else that has even a rudimentary ability to sense radiation, are responsible for our material world. When an electron becomes aware of the field of a proton, they will be close enough together to be considered a dimension of space, and an element of matter (hydrogen gas).

You may wonder why physics consider quantum points as *information*. The reason is that whenever a particular quantum wave shape is phased and amplified to become a material *something*, it becomes an electron, proton, neutron or photon, depending on the information contained in that wave. Some of these *things* require more than one point. It is obvious that these points (phased waves) contain some kind of coded information that when they are put together they become one or the other of the basic points that then combine to make an atom of something. It is also obvious that this coding system is much like that used in our computer systems. The 1 and 0 binary code that you see on a sheet of paper is in reality a + and - system in your electronic machines. That is why quantum physicists equate the quantum dimension with a universal intelligence. All of the waves in that quantum system are negative, positive or neutral energies which convert to physical points with the help of a consciousness. In conjunction with one another they are the basis of matter as well as spirit.

Transportation of information from one end of our universe to the other instantaneously could be accomplished by the virtual particles that fill our space. When these particles enter our material universe, they react to the magnetic moment of any material particle that they appear near. This then becomes information that the virtual particle returns to the quantum dimension with. Since the quantum dimension has no time or space, that information is available anywhere in that universe immediately. The same is true in reverse. A quantum virtual particle has a connection to all of the information in the quantum realm. When it enters an area that is filled with material particles (such as our brain) those particles will have access to that quantum information if we develop the talent to extract it. Bells theorem *demands* a superluminal connection from one end of our universe to the other. This experiment has been done many times by measuring the phase of diverging photons, and is accepted by quantum physicists. All matter in the end is quantum material (information only) thus all is interconnected.

It seems a shame to me that most of the research I am aware of about consciousness is restricted to physical manifestations. If you trace neurons and electrical and chemical changes in the brain, it is inevitable that the experimental outcome will be materialistic. All of the science that we have been reviewing on these pages indicates that this material world is merely an extension into space-time of a widely interconnected entity. Consciousness isn't restricted to our material universe.

Many neurologists and neuro-scientists believe that we use only about five to ten percent of our available brainpower. Inquisitive individuals are aware that nature has never evolved a system that isn't useful in some way. Ninety percent of our abilities are waiting for us to come out of our own dark ages. Or perhaps that ninety percent is being used for things that we aren't aware of.

Simple procedures of physical and mental relaxation will allow you to change your level of consciousness from a high level of awareness of your physical surroundings to one of almost a complete disregard of things physical. It is through this procedure that you can begin to be more aware of the spiritual (quantum) information that surrounds us.

Most of us tend to misuse the word *real*. A dream is real to a dreamer. A vision seen in delirium is real to that person. Someone talented in seeing a body's aura sees something real to them. Reality is a very personal thing. These things that we sense are the result of an immense number of quantum particles/waves moving through the brain. The brain/mind will compare this coded information to information of the past and will re-cognize it for what it is.

Human misconceptions have so distorted the spiritual aspect of our existence that much of the misery of our present is directly the result of prejudice of variations of belief. The North African and Near East problem are an example of religious intolerance. It is abundantly clear that we are all made of the same materials. We also all exist in dimensions other than the space-time we are so aware of. Bells theorem states that quantum material things once they have been in contact physically are forever linked in the quantum (spiritual) world.

The science of biology tells us that we are all related genetically and that means all the way down the chain of life. Our bodies are constructed of matter put together in the Big Bang, as well as the centers of nova producing stars.

It seems to be true that we are materially universal entities, (matter appears universal). I find it very likely that we are also spiritually universal entities. We are making progress emerging from our intellectual darkness. We are making less progress emerging from our spiritual ignorance. The wars of our world that are fought in the name of one religion or another are a good example.

VIRTUAL PARTICLES AND OTHER WEIRDNESS

Before I go any farther, this seems to be a good place to discuss a little more the quantum phenomena of "Virtual Particles." The title *virtual particles* are misleading to some. We see in other areas of this text that *particle* is used mainly as a word to locate a field of influence. An electron is referred to as a particle when in reality it is a mathematical center of quantum processes. It is true therefore that a virtual particle is almost the same as a true particle. The term was chosen because of the quickness of their arrival and departure. They arrive from nowhere and depart to the same place so rapidly that the rules of physics demand that they not be considered as a part of our physical world and do not obey our physical rules.

The literature of quantum science uses the terms that I just said "come from nowhere and depart to nowhere" with tongue in cheek. It is evident that these points of influence boil up from the quantum dimension (in pairs). The reason for this I believe is that in the infinite collection of quantum waves, it is probable that similar wave shapes will drift together and increase their amplitude (by accident) just enough to allow their peaks to intrude into our physical dimension. Quantum physics tell us that space really consists of foaming activity. They are as yet unsure what the foam is. Many believe that it is a collection of other dimensions all rolled up tightly and that we are incapable of measuring them. I stick to my belief that it is a natural activity in the quantum dimension.

Physicists are aware however that these particles do have some ill-defined influence on our physical as well as our quantum states. If you take an area of space (a vacuum jar) and evacuate all particles such as air, dust and everything physical, you will still have a jar full of virtual particles. This is true because they are not considered as part of our physical world and do not obey our physical rules.

It is the quantum uncertainty principal that allows virtual particles to emerge briefly from nothing at all, and then simply disappear. Remember that the quantum state can allow for an infinite choice of functions.

The quantum world transcends time and space. Virtual particles have instant access to everywhere and every when. Once again those *particles* are expressing possibility. You will see in our text that a pressure exerted by our consciousness influences the selection of one possibility, and changes it to a probability and continuing to a reality.

Physicists are aware that even though the impression is very short, there is always a trace of their comings and goings. They measure the amount that a physical particle is displaced by the magnetic moment of the virtual particle. This is true even if they cannot be directly observed.

For many years the belief was that electromagnetic waves needed a medium through which they could propagate. This invisible medium was called the "Ether." The science of the time demanded that this medium be a physical thing. Experiments to prove the existence of the ether served only to show that there was no such thing.

As we discuss virtual particles and the quantum dimension, you will notice that their quantum effects really could be a non-physical channel for the wave information of Nick Herbert's "quantum stuff". They could also be the channels for all of the so-called *psychic phenomena* that we observe and misunderstand.

Quantum, mechanics (the science) allows energy to emerge from the nowhere of quantum land, as long as it goes away again real quickly. Since matter and energy are interchangeable, the result is the emergence of particles very briefly out of nothing. This has required a definite change in our meaning of *empty* space. The vacuum of space as well as our atmosphere is actually teeming with virtual particles. The spaces between the atoms of all ordinary matter also are jammed with these particles that appear, interact and disappear. They are immediately replaced by other virtual particles. This interaction isn't just a theorist's speculation. There have been real effects detected in laboratory experiments.

The physical laws of the universe require that energy and matter be kept in a very tight balance. This can be seen by looking at Einstein's equivalency equation. It is also the reason that our universe has the velocity limit of light speed. Without this standard matter would be chaotic and mostly non-existent. The uncertainty of the quantum state allows particles to appear for a *minute* amount of time without having to make adjustments universe wide. It is true that material particles can be made if we supply enough energy from our universal supply. Particle accelerators create new particles all the time by supplying energy from some other source. This transfer maintains the balance that nature demands. These particles that appear when protons are smashed together are new particles created from the energy of the protons in motion. There is no need for any universal adjustment since no energy was borrowed from the quantum dimension.

Scientists are aware of the arrival and departure of these virtual particles, they are also aware that they *do* interact with our material particles. The "why" of this activity is still a mystery to them. (?) Although few experimenters will say in print that these particles come to us from that quantum dimension, if you read between the lines of their experimental proceedures this is the home of these ghost like particles.

It is *my* opinion that these points of information act as a conduit to that fifth quantum dimension. All through this book I will equate that fifth dimension as a super-consciousness and also possibly the material of the "Mind of God."

All of the experimenters tell us that the quantum dimension is one of information. Our limited consciousness is also one of information. Since the quantum virtual particles are in the tissues of our brain in great numbers, it seems that our thoughts and activities are impressed instantly in that area of quantum information.

At the same time it seems that if we learn the technique of altering our state of awareness we can extract the information that is part of that super consciousness.

During my working life as a therapist, I came in contact with many people that practiced various forms of psychic phenomena. These people *always* altered their state of consciousness by varying degrees, and some were not aware that they did so. By using an EEG machine it is possible to monitor their brain waves, and they always showed an increase of the alpha wave that signifies relaxation and a search of internal information. This internal information in most cases was not from their experience, but from (I believe) the knowledge of the super conscious (quantum information). Many physicists tell us that the empty space between planets and galaxies is filled with foam. That's the word they use. My bet is that foam is merely the way the tremendous amount of virtual particles coming and going appear.

Physics has become kind of an eerie place, even for a lot of professional physicists. Because our senses tell us that we bump into things, most people like to think of things as made out of solid material. With the advent of quantum physics, that solid material of the past has turned into quantum *stuff.* To make maters worse, even electrons are no longer things. Physics has become (like it or not) a branch of philosophy. In our philosophical search of the environment we inhabit, we have spoken much of the electrons. We discus very briefly the Hydrogen atom, which was the significant element (75%) manufactured in the Big Bang.

The hydrogen atom is composed of one proton and one electron. The proton manifests as a positive point, and the electron as a negative. Because these charges in the atom balance each other, the atom as a whole has no charge and is very stable. This atom would be classified as a material particle since it does have a small mass and it does occupy our space. Even though we would call these atoms *things*, they are still connected intimately and instantly to all other atoms in the universe. It is holographic information. They are part of a quantum system that is universe wide. The writings of mystics and physicists are coming closer and closer together. You have only to read books like "The Tao of Physics', by Capra and "The Dancing Wu Li Masters", by Zukav to realize this. This irritates many physicists, but some have learned to be open to the information.

Physical experiments show that not only do we influence our reality, but we may also actually create it. Our reality seems to be what we choose to make it. We do this with all of the forces and entities around us. It seems very natural for us to assume that our consciousness resides within our body. We are *in here* and everything else is *out there.*

Because of the intimate connection between our body and brain and the virtual particles that move back and forth between our space and the quantum dimension, we are a source of information for that super conscious entity. Since it is true that we are so closely connected to those virtual particles and the super consciousness, it is also true that the study of quantum physics is also a study of *our* consciousness. The most difficult part for us to understand is the makeup of our body and the other materials of our universe. Experimentally it is a fact that the materials of our body and all of the rest of matter in our world are not really native to our space time dimension. They are quantum waves that are changed into matter by the collective consciousness of entities that are at present residing in space-time. To look at it another way, each spiritual entity exerts conscious pressure (force) on the quantum information to raise its amplitude high enough to cause its peak to become real material in our universe. These material points will then arrange themselves in a pattern that will define our body space. Each body is a separate universe residing in the larger universe. Each of these has its own basic oscillation that is often detectable by psychically sensitive people who practice psychometry. Initially a spirit outside of space-time controls this body building. You are aware that a unit of space can be any size, and that there are spaces within spaces. When you occupy an automobile it can be a small universe for you and a fly. When the fly moves from the back to the front, it moves at the usual fly rate, even though your universe is moving through another universe at 60mph. The space your consciousness provides for you to experience events in our material world, is your body.

Since our world is four dimensional, the quantum points can arrange themselves in any way. They get their instructions from another source of information called your DNA.

All we can do at present is to speculate about what we would see if we could view an electron, or any quantum point.

In my mind, a quantum point viewed from our space-time realm would look very much like a total eclipses of the sun. This would be a very dark spot wit a corona of light around it. Within the dark area would be a network of that same light that shines around the outside. This light is an indicator of the force field that is present as well as the spiritual presence.

This picture is very similar to what you would see from a safe distance in space looking at a black hole. The black hole would have an event horizon beyond which the laws of physics do not apply. The halo around a black hole and a misty kind of light across the entire object would be from the evaporation of material within. An electron would be much the same since according to our laws of physics a quantum particle is not a part of our space. There must be something similar to an event horizon masking our dimensions from another. If you could safely enter a black hole, you would probably exit into another part of our universe, or into a different universe altogether.

Else ware in the book I speculate that the quantum particles are all wormholes to other dimensions. This is for want of a better description, but they *do* connect for sure to the quantum dimension. If this is so, our essence or true being is of another place. I call this other place our spiritual home, and also the quantum dimension of information.

While a black hole derives its power to connect dimensions from its huge mass (energy), a warp of a different kind would power a quantum point. The power of consciousness keeps us and everything material connected to both universes at all times.

Try to picture your body as billions and billions of ports to other dimensions. The other ends of these ports open to a dimension or dimensions that are pure information. Think about your material life for a moment. *Everything* that you do or sense can be reduced to a binary code just as it is in your computer, with the addition of emotions etc.

The binary information of the quantum dimension can be available to our dimension with the aid of a properly directed thought. The quantum waves that compose all of this information need to be amplified by phasing the waves so that they add and increase their amplitude. This action will increase them from the *possible* to the *probable* and to *reality.*

Our thought processes increase the activity in various areas of our brain and influence the shape and amplitude of the quantum waves (coded information). This directed activity can produce material things, or in laboratory experiments give singular answers to specific questions.

As a therapist who uses hypnosis as an aid, I have witnessed amazing changes in people through their directed attention.

Waves, Fields and our Spirit

We are intensely aware of our physical body because of the sensations it receives, both good and bad. We have in the past assumed that we also have a spiritual body because the religions of the world tell us so. Since the era of quantum experimentation, we now have reason to believe that science can demonstrate our spiritual presence with experimental evidence. The dual existence of our physical/spiritual particles seems to be the evidence we have been searching for all of the time.

Quantum theory says that before we measure a particle, it isn't really a particle it is in fact a wave of possibility. This wave is spread out over a dimension that is more than space-time. It is in fact space less and time-less. An unmeasured quantum resides in a less restricted environment than a measured quantum. Unmeasured it is part of the information of the quantum dimension and is free to be everything. Measured, it must conform and be a part of our physical world. Unmeasured it can be everywhere and every when. Measured it must be here and now.

If an experiment is assembled to measure the velocity of an electron, that answer is allowed. If an experiment is assembled to measure the velocity and also the position of an electron or any quantum, one measurement will be precise and the other only a range of possibilities. There seems to be a rule of nature that if you know where it is, you can't know where it is going. If you know its destination you can't know how long it will take.

It may be good to remember that the fuzziness of the quantum state (unmeasured) can be ascribed to the special dimension in which it resides. The unmeasured quantum exists outside of space-time. It is for this reason that all things are possible for quantum articles (including you and me) are present once. Remember also that the quantum dimension consists totally of information. This information must be assembled by a consciousness, and the waves increased in amplitude by phasing them together so that they add. This enlarged waveform then will have the tip of the wave existing as a physical particle in our material world. The rest of the wave that is below that very high amplitude becomes part of our spiritual body. They must exist together.

Quantum particles that become material particles are always accompanied by their quantum (spiritual) basis. It is for this reason that when we talk about our body, we are talking about two bodies. One resides in our material space-time world and the other still resides in the non-material quantum (spiritual world). I call it spiritual world because it is how we have come to think of a spirit. It is non-material and exists in a dimension that is timeless and space less. It is however located in the same material place as our matter body as long as the consciousness of the individual continues in that material body. It is our immortal consciousness and will that selects and assembles the quantum waves that will become the material body we think of as "I".

Keep in mind that even the tissues of our brain are ultimately quantum materials. The communication that occurs in these tissues is electrical (electrons). Our physical senses have developed to give us information about our material habitat. It appears that our body is being used very much like a satellite in our space-time world. We build machines and send them off to other worlds. We place sensors aboard to send us information about the conditions in an alien world.

We talk a lot about the electron simply because it is the most important quantum component for material things. The electron in combination with the neutron and proton, decide whether a molecule of matter will be lead, or gold. Various combinations of electrons and nucleus components give us all of the elements of the periodic table. These things in turn combine to form all of the matter of the universe.

Because of all of this talk about the electron as a particle *and* a wave, our brain strains to picture something that can be both at once. Our physical science requires that we define things in material ways. The particle and wave are mental pictures that we can understand. A field, such as gravity is a bit more difficult to understand. Never the less, physicists have assigned a particle called the graviton to this field, so that we can more easily visualize the transfer of force. A transfer of force is also required between the fifth dimension (the quantum realm) and our limited four-dimensional space. I am convinced that the unnamed field that connects this world with the world of possibilities is intimately attuned with the virtual particles that are not at present considered as part of our space.

Heisenberg discovered that the imprecision that is always present during experiments to discover the momentum and position of a particle, has the inverse relationship that we have talked about. Also that this relationship is true in every imaginable experiment involving the simultaneous measurement of pairs of particles from the same source. This uncertainty principal is a challenge to causality. In our scientific studies we have been taught (in the past) that every effect is preceded by a unique cause. The quantum uncertainty will not allow a prediction of an event in the future. In classical physics, if we know the present in every detail, we can calculate the future. This is true only for uncomplicated things. It doesn't work that way in quantum physics. Any action can have an infinite number of reactions.

The future for a quantum particle is hazy at best, and larger quantum entities (you and I) have the same long list of possibilities. The question, "what is reality?" is still a priority for mystery loving physicists. Are things really there, or do we with our conscious observation create them? Nick Herbert in "Quantum Reality" says:

> "One of the main quantum facts of life is that we
> radically change whatever we observe. Legendary
> King Midas never knew the feel of silk or the human
> Hand after everything he touched turned to gold.
> Humans are stuck in a similar Midas like predicament.
> We can't directly experience the true texture of reality,
> Because everything we touch turns to matter."

Einstein proved to us that reality is subjective. Time, for instance is relative to your velocity or mass. Someone or something at a different velocity or part of a greater mass is subject to a different time. In effect you carry your own universe with you. This is the basis of the old story about the fly moving from the back seat of your car to the front. Your universe is his, and he moves at the usual fly rate. It is just as proper to say that your universe is at rest and the ground is moving by at sixty miles per hour. We have to remember that nothing physical in our universe is stationary (at rest).

The empire state building is in rapid motion as the earth turns on its axis. Our earth moves in its orbit about the sun. Our Milky Way galaxy is moving away from all other galaxies (with the possible exception of one)

I mention all of this just to remind you that each center of consciousness (yours and mine) in this physical world, live our lives in various stages of isolation. This is true in our world of matter, but does not appear to be true in the world of spirit.

When we have finished with our body, because it is worn out, or because it has been injured in some way, we reenter the quantum spirit world where everything is one unit. We will be able to experience true togetherness with all.

ACTION AT A DISTANCE

Professional weather forecasters sometimes say things like "if someone sneezes in Africa it can cause a hurricane in Florida." What they mean is that a tiny movement of air in one place on the globe can have a devastating effect some time later in another area. There is a whole branch of science dedicated to research into the effects of Chaos. The mathematics of this field is overwhelming to all but the experts; consequently progress is slow toward understanding. To be aware of all of the negative and positive pressures and waves is tough enough, but the way they add here and subtract there is frustrating. In weather as well as many other fields it is apparent that complex but ordered results can come from chaotic circumstances. I'm sure that with enough reliable information, weather prediction will get better and better.

My point is that an event quite distant from you and I can effect us sometimes instantly and often at some future date.

Some years ago Sheila Ostrander and Lynn Schroeder published "Psychic Discoveries Behind The Iron Curtain." They reported on several areas of psychic research going on in the USSR at that time. One of their reports was an experiment supposedly carried out with the help of the Russian Navy.

The experimenters selected a doe rabbit that had a new litter, and placed her in a laboratory. She was then wired to instruments that would monitor and record physiological changes in her body. The young rabbits were placed in a submarine and taken to a great depth in the Atlantic. From time to time, one of the young rabbits was killed. The time of execution was noted from a clock that was synchronized with one in the laboratory in the USSR.

At the conclusion of the experiment, there was strong evidence that the doe knew something was amiss with her babies. Each time one was killed there was a corresponding instantaneous physical reaction in the mother rabbit.

If we can believe that this experiment was controlled properly, it does seem to indicate that there is an extended link between mother and offspring, at least in rabbits. I would like to believe the same is true for all sentient beings. In later chapters we will discuss how this link is at the same time a physical one and a psychic one. Laboratory experimentation has shown that each quantum particle is in constant contact with all others. These experiments use paired electrons that are diverted in opposite directions, and they still maintain their pair connections.

The Russian experimenters indicated that the influence was noted instantly in the mother rabbit. This was no doubt deduced by comparing the time of action on the boat, with the time of reaction in the laboratory ashore. This seemingly instantaneous transfer of information doesn't rule out physical reception of signals. Many radiations that we work with every day travel at or near the speed of light. At 186,000 miles per second, a physical signal could go around the world many times before a delay would be noticed.

Connecting signals between physical entities have been known to physicists for some time. In later pages we will discuss some of them. For now it is sufficient to say that electrons confined to relatively small areas seem to remain linked together later on no matter how far apart they become.

Remember that our bodies, our cells and our atoms are all combinations of particles. The electron happens to be a good possibility as an information messenger.

We are hopefully all aware of the chemical mechanism of sexual reproduction. The male and the female each provide one half of the DNA chain necessary to produce a new human. Remember that these tiny bits of elemental material have resided with the mother and father for some time. Remember also that each of us has a personal overall electrical oscillation. This comes from a combination of all of the materials (quantity and type) of our body. This body field is as individual as our fingerprints. Our electrical oscillation is imprinted on and is carried by the DNA that combines to form a new individual. This physical field is in addition to its unit electrons.

Just as the child has characteristics of both parents, he/she is also sensitive to their individual body fields.

In Quantum physics the property of locality arises. In most cases, local forces control the behavior of local systems. Forces such as gravity, light, heat and so forth greatly effect the growth of a fruit tree. They also affect us in similar ways. Most of what we sense is because of these forces. There are times though that non-local forces become involved.

It has been found when experimenting with quantum particles, that these sub-atomic particles can interact in one locality and then move very far apart. The rule for quantum particles is that even though they end up on opposite sides of an immense space (even the universe), they are still part of a unit and must be treated as such. Measurements performed on one particle will mirror the status of another particle of the same unit. The behavior of any one particle is measurably tied to all of the others no matter how great the distance between. This rule of the quantum that has been experimentally tested many times would seem to be operational in the rabbit test mentioned earlier.

Einstein disliked this finding because of its *action at a distance*. Recent experiments however have confirmed that these non-local effects do occur.

There is a bit of mother and father in every offspring. It seems to me that the result of the Russian experiment should have been expected. The problem for some experimenters is that this result seems to be categorized as a psychic phenomenon. As we move farther into this book, you will find that many so called psychic happenings are merely misunderstood quantum events.

In order to understand more easily the phenomena of *action at a distance* and other equally strange quantum events, it will be necessary to briefly discuss time and space.

Our Four Dimensions

Earlier we discussed how we could build space simply by indicating with quantum points the edges of our space. Essentially nature (universal consciousness) has done the same thing. Present thinking among the physical scientists is that without matter there is no space. Matter in this case is considered to be anything composed of electrons, protons and neutrons. These are quantum entities, and can properly be considered as waves of possibility.

Space as we considered it before is just a static thing. Space without time is very boring. There is no movement or change of any sort. The addition of time however makes a great difference. Time allows galaxies to form and stars to burn. The expansion of the universe and the explosion of the super novas are all because of time. Material drifting from these explosions has allowed our planet to form, and biological units to move up the scale of life.

Physicists consider the photon as a messenger particle. Whenever elements change by adding or losing electrons, photons are employed. The same is true if an electron moves to a higher orbit (gains energy) or drops down to a lower orbit (loses energy). I consider the photon to be a particle of time. Photons always travel at the same velocity. They don't get a boost from the platform they leave. If a spaceship were traveling at half the speed of light, and emitted radiation in the same direction, the radiation would still be measured at 186,000 miles per second. Einstein's works illustrated the strange effects encountered as speeds approach the speed of light.

By adding the dimension of time to the three of space our consciousness is given the ability to evolve. Without change there is no evolution. Throughout the book we will talk about the timeless space of the quantum dimensions. In these dimensions it is possible to immerse consciousness (yours) into the flow of time without affecting its course.

Just as a grandfather's clock needs the swing of a pendulum and a quartz crystal watch needs a vibrating crystal, the universal timepiece (nature) needs a constant speed. The relationship between matter and time must be universal or the construction of matter would be chaotic. Time isn't a human construct, but is necessary throughout the universe, and dictates the construction of matter. Space-time dictates not only the orbits of planets, but also the so called orbits of the electron.

Many of my lectures in the past were for the sole purpose of getting people to think. Most Americans (not all) lead a lifestyle that is not conducive to individual questioning. This is so especially in the areas of science that bring into question things that have been left mostly to the various religions. Control of *the flock* could get difficult when questions arise. Many religions try to impede the flow of scientific information without recognizing that the more we experiment, the more it seems there really is an intelligence behind the universe.

Back in the 1700's and early 1800's it took weeks and months to travel from the east coast to the west coast. After the railroad was established the trip was reduced to a week or more. With the advent of jet aircraft only hours were necessary. Are these areas getting closer together? Our physical senses say they are, but intellectually we know they are not. If at sometime in the future we learn how to broadcast our bodies from place to place like they do in the "Star Trek" series, does that mean the east coast and west coasts occupy the same space? They will seem to, since the trip from one to the other will be relatively instantaneous.

We have a long way to go before we can do anything like that but each step along the way is going to require a change in our thought process.

Einstein's general and special theories of relativity made it quite evident that our ideas about time and space need revising. In the past, we were content to believe that space consisted of three dimensions. We named them length, width and depth. Time was considered a separate phenomenon, even though we calculated distances with references to time. It took the good doctor Einstein's writings and mathematics to convince some to modify space and time to space-time. Good science requires a lot of good ideas, data and facts. Without these it would be difficult to approach any scientific truth.

THE FIFTH DIMENSION

I will have to test your ability to visualize while I explain a dimension other than the four we are used to.

Scientists that discuss the Big Bang and its aftermath like to say that you can't picture our universe expanding into anything because our space is all there is. At the same time they should all be aware that quantum scientists are able to describe a dimension that is apparently composed of coded information. This quantum dimension is at the same time external to our four dimensions but also is intermingled with them continuously. Consequently there *are* more dimensions than our four. The exact numbers of dimensions are still under study but they seem to think there are at least ten to twelve.

Let us begin by visualizing our four dimensions (three of space along with time) as a bubble (one unit) expanding in a fluid. This fluid can be that quantum dimension that physicists speak of. Recognize that the skin of that bubble is composed of quantum particles. These particles have a kind of dual citizenship, they belong to another dimension (call it the 5th or quantum) but they influence and delineate space-time.

The hypothetical fluid I spoke of earlier is the actual composition of this other dimension. We know it is there because the virtual particles scientists speculate about come from there and go back to the same place so rapidly that they seem to form sort of a foam in our space. Since at present it is impossible for us to know for sure, its total composition is up for conjecture. The quantum scientists and I see it as a place of information (like a consciousness). One way to define a consciousness is an awareness of change. The field of this other dimension can be the virtual particles that we see moving back and forth between the other dimension and ours. We become aware of the information that it contains through the possibility waves of the quantum (5th) dimension.

Space can be any size, and there are spaces within space. We began this chapter by visualizing a bubble expanding into something else. Our problem though isn't quite that easy. The expanding bubble is a good representation of our space-time universe, but we also know that those quantum particles that we see as the skin of the bubble, (edge of the universe) are also the skin (outline) of our body space as well as all of our organs and brain. This carries down even to the cells of all of our systems. Our bodies are in direct contact with the quantum intelligence just as is the edge of the universe. It is sometimes easier to see our space-time as a multitude of universes, all of which are surrounded and penetrated by the possibility of everything. *The quantum entity*, this realm of possibility the physicists speak of always, could be the collective unconscious that Carl Jung spoke of. On the other hand it could be the God that religions speak of, even though the religions differ greatly.

Quantum possibility is the essence of consciousness as well as the material of the 5th dimension. If we pay attention to the writings of our physicists, it becomes more and more obvious that we truly exist in many dimensions. Size and mass are areas of change between our normal dimensions and those we think of as spiritual or extra dimensional. At the Plank scale (quantum sized) time and space break down into an infinite amount of alternatives. This is the possibility wave of quantum mechanics. As you move upscale in size and mass, each entity whether it is an atom or a human body has a time that occurs at its own rate. Many experiments have been done to show that as you increase mass, time slows. This phenomenon occurs with everything.

A basic of physics is that in the absence of matter (mass) light travels along a null surface (straight line). If a small amount of mass is encountered in this light trajectory, relativity says that the path of the photon is bent or curved around this mass by a measurable amount. If the mass is great enough, such as a very large collapsed neutron star, approaching matter (photons included) will be captured and not be able to exit again. The same is true of photons attempting to exit the star from inside. This then is a black hole. Since our laws of space and time depend on the velocity of light, the absence of this measurement effectively the area from our universe as well as the space-time within the singularity. We are aware of the presence of this entity (the black hole) because of its effects on our universe, but the internal workings of the thing are unknown to us. What I have just said should remind you of the quantum point we talk about throughout the pages of this book. We are aware of quantum points and virtual particles because of their effects in our material world.

A single quantum point would be unnoticed by us because it really is not of this world. The single point does not constitute a dimension. If we add one more point, a proton perhaps and make it an atom of Hydrogen, we still wouldn't notice it unless we had billions of them. These atoms are now however a part of our space because of mass and the dimensions between the points. Part of your change of attitude about your immortality requires that you understand that there is a change in the relative rate of time around all bits of matter that is proportional to the mass. If you are pure spirit, there is no mass, and consequently no time. The word mass can be interchanged with "the number of quantum points." In our material world the change of time rate is unnoticed within the shell that surrounds the mass. Gravity is the force that dictates the slowing of time within the shell.

Physicists for some reason have told us to visualize the space-time distortion that is due to mass as gravity *well*. This mental picture would indicate a depression in a surface of space (a flat sheet) that is a product of the mass that is present. To be more able to recognize that each bit of mass is a universe in itself, we must visualize space-time as four dimensional when we speak of the effect of gravity. The picture of a dent in space-time is incomplete. In a space-time universe, the introduction of matter (mass) will construct a spherical shell around this material. Our scientists describe this phenomena quite well when speaking of a black hole, but neglect to visualize the same type of spherical shell around a planet, or even a human body. Within these shells time is at the rate dictated by the mass. A gnat flies around in our earthly time scale, but the internal workings of the gnat's body are geared to his body mass. Because of our greater ability to process information we are aware of these other time scales.

If we could somehow accelerate a spaceship to near light speed and still orbit near our earth, we would view the activity on the planet as a blur of motion. Our time scale (on the ship) would be very slow, and years would go by on earth like the snap of a finger.

Dr Einstein told us that time was relative to velocity or conversely to mass. If when we die we leave our material body behind and occupy only the spiritual (quantum) body, we enter a realm in which time has no meaning. The quantum waves of information have no material component and as a result have no space or time restrictions. It seems as though in our spiritual condition we could be where ever or when ever we wish. What a great chance that would be to relive the experiences of our material life (all of them) and once again ponder our mistakes and plan a new lifetime to correct our shortcomings.

I believe, as do many, that our God (super conscious) isn't judgmental, but is interested in our evolution toward a higher state of being. Our evolution would be advantages to the consciousness we call God, and would at the same time increase its evolution.

THE NON-LOCALITY
OF THE SPIRIT WORLD

It is necessary that I expand a bit on some of the information that I have expressed previously.

John Stewart Bell in his experiments was able to show that an ordinary model of reality as well as a contextual model, has to be non-local (space less). In local reality it is necessary to state that influences can't travel faster than light. Bells theorem argues that in any reality of this sort (ordinary matter); information could not travel fast enough to explain the quantum facts. Bell assumed therefore that reality must be non-local. The theorem requires that influences and information that are not local connect its objects (quantum stuff). Bell states that a faster than light communication for ordinary states is a necessity to explain quantum facts.

Bell has proven to all quantum scientists that the two models of reality, ordinary and contextual must also include signals that travel faster than light. If this is so, our reality consists of more than the four dimensions we are acquainted with.

At this point it is necessary for me to review some of the elemental features of electrons, so that we can continue to discuss our material connections with other dimensions. A very simple experiment using an electron gun (like the gun in your picture tube), a barrier with a tiny hole in it, and a phosphor screen will give plenty of evidence of the electrons particle nature. With one hole in the barrier and a proper adjustment of the brightness (energy level), you can see the electrons striking the screen one by one as a particle should. If we were to put another hole in the barrier (on the same horizontal line), we would begin to see bars of electron hits on the screen. Some of these bars would be of a light shade, and others would be darker. This interference pattern is what would be expected with a wave output from the gun. This is the same gun we used before but the barrier now has two holes. If we put a piece of tape over the second hole, the pattern will go back to a particle indication. That electron knows that there is only one hole when it leaves the gun. The Copenhagen interpretation of quantum law says that reality is created by observation. Waves striking two holes should create interference patterns, so it does. Just moments before, the same gun was only putting out individual electrons and no waves. If you feel frustrated by this weirdness, don't feel bad because Dr. Einstein didn't like it either.

Most feel that an electron is a point particle. That is a particle whose real dimensions are zero. It is difficult to visualize a particle with no dimensions at all, but it works mathematically. All of the rest of the elementary particles seem to be point particles also. Atoms however begin to show more definite size. In this case think about a point particle. It has no size (no dimensions). Two points however will define a dimension (length) and will indicate space as well as time. An atom of Hydrogen (one electron and one proton) will occupy space in our space-time dimension. Electrons by themselves seem to be pure quantum stuff. By that I mean all field focused to a point. Photons, neutrons quarks and all elementary particles are likewise quantum stuff. They seem to be waves of information radiating from a mathematical point. This makes sense since they are all native to the quantum dimension that is composed totally of information. Some physicists believe that objects like coffee cups, trucks and planets would exhibit quantum wave properties under the right conditions. Limitations of wavelength and instrumentation make it impossible to observe. It seems that elementary particles behave like particles if we watch them. If we aren't watching they behave like waves. This is the part of quantum physics that bothers us.

Since our bodies are composed of material cycles, that are themselves the peak amplitude of quantum waves of information, we can consider all of the information below the peak as quantum (spiritual) non-local information. We walk around with two bodies. One, our material (space-time) body, and the second one that is totally enmeshed with the material body, but is at the same time our spiritual (non-material) body. Bells experiments require these two conditions of our material and non-material points.

Super Luminal (Faster Than Light) Connections

In order for us to make sense out of some of the psychic events that occur mostly in a random way, it is necessary that we have a basic understanding of the two different universes into which we are immersed.

Classical Newtonian physics speaks of matter and fields. The connection and interplay of these two entities form the universe we have taken so much for granted. Only in the past sixty or seventy years have we begun to question how firm the ground is that we stand on.

Probably a large portion of the educated population tends to think of matter as little balls of *stuff*, packed together to form various *things*. Quantum theory has forced a change in that thought process. Over the years classical physics has come to recognize four fields. They are electromagnetic, gravitational, strong and weak fields. A field is defined as a distribution of forces in space. These fields all have limits in their ability to influence matter.

All mater in the universe is imbedded in electro-magnetic and gravitational fields. It has been found that if you move matter, be it electrons or a planet, the motion causes waves in the fields. James Maxwell determined that the velocity of these waves depends on the electro-magnetic forces. ///The influence of electricity, magnetism (really one field) and gravity are reduced as the distance squared. This law of reduction of force mathematically never goes to zero. Technically their range is infinite. On the other hand the strong and weak forces are confined to the atomic nucleus. As research progressed, physicist James Maxwell proved that the electric and magnetic fields were two aspects of the same field.

All matter in the universe is imbedded in electro-magnetic and gravitational fields. If you move matter, be it electrons or planets, the motion causes waves in the fields. Maxwell determined that the velocity of these waves depends on the electromagnetic forces. When the velocity turned out to be exactly the same as the speed of light, it became obvious that *light* was really very high frequency electromagnetic waves.

It will be good to remember that motion of a charged particle, whether it is a planet or an electron, will affect its surrounding fields. Also keep in the back of your mind that a "wave" is a rotation of polarity from negative too positive. This alteration can be rapid or slow (frequency), and it can be large or small (amplitude).

There remained a puzzle in physics as to the nature of matter. The mathematics of Einstein and work in the labs has shown that light has particle properties. The distinction between field and matter was becoming less and less.

This blurring of distinction between field and matter tends to indicate that the universe is composed of one substance that is masquerading in various costumes.

Nick Herbert in "Quantum Reality," calls this "Quantum stuff." In order for this quantum stuff to exhibit the qualities of wave and particle at the same time, it would require that we visualize an extension into other dimensions. It is difficult for us to visualize time as another dimension, and it is even more difficult to visualize a fifth dimension.

Our short discussion of the construction of space showed that those three dimensions are all at right angles to one another. I may be alone in my belief that this relationship changes with time and all following dimensions.

Some philosophers feel satisfied and can relate to quantum stuff. But it is disturbing to physicists. It's like trying to catch a ghost. If you reach out for it, there is nothing there.

It helps somewhat if you visualize a point (something that has no physical dimension) alternating around a neutral center in positive and negative directions. A wave of course has to alternate. It doesn't have to be a physical movement but it must be a change of potential. The center of the alternation (the null point) could correspond to the particle aspect. Most people are aware that an alternation doesn't have to begin at zero. You can alternate around any potential, so the particle (null point) can represent any value.

In later pages we will discuss how an electron (particle) can cause a wave aspect in a medium of virtual particles. These are pairs of quantum points that permeate all of space. According to the dominant theory, they appear and disappear too quickly to be recorded as part of this universe. There is speculation however that these virtual particles have a function in energy transfer in our space-time universe.

Heisenberg's matrix system, Schrodinger's wave mechanics and Dirac's vector system all are used to describe the wave characteristics of quantum stuff. It is my belief that more progress will be made when consideration is given virtual particles and their alteration between dimensions.

The conflict between classical physics and quantum physics is a philosophical and emotional difference. Is matter really there or isn't it? When we stub our toe on the table leg, do we hit something tangible or is it an intense field? The big names in quantum physics (Bohr, Heisenberg, Bohme, and Von Newman.

ETC) all agree that electrons are not things. They are merely an intense field that masquerades as something solid. Years of good science have proved beyond doubt that all elements are composed of various combinations of electrons, protons and neutrons. If these combinations don't qualify as things, what do we have? We have no things, commonly called nothing.

The reality question in quantum physics really isn't about the actual existence of quantum points, but how they manifest their attributes. Quantum theory proposes that the world/ universe isn't made up of solid physical bits of something. The theory contends that all quantum particle/waves, including the electron, do not possess all of the attributes innately that we would consider necessary to form solid material. They do possess mass, spin and charge that will give them identity from all other quantum entities. These values are the same for all electrons under all conditions of measurement. This would seem to qualify them as ordinary objects. As we have discussed before however the dynamic attributes of position and momentum seems not to be attached to the electron without other qualifications. The way an electron acquires these dynamic attributes is determined by the mode of measurement. This curiosity is the basis of the reality question. If these attributes change or defy measurement, there is no basis to call them ordinary objects. It seems that a portion of the electrons' attributes belongs jointly to the electron and to the consciousness doing the measuring.

Apparently the electron has no solid value of position or momentum until you measure for one or the other, but never for both at the same time. If you manage to obtain a good measurement of its position, the momentum is indistinct and visa versa. Nature will allow us to be only partly aware of the attributes of the electron at any one instant of time. It also seems that if you remove the measuring device (which is ultimately a consciousness), then both position and momentum are hazy and arbitrary. Contemplation of this state of affairs brings to mind holographic images that seem real, but aren't there to touch. We know better though because the whole universe is constructed of these particles/waves and we can touch them. Can't we?

In an attempt to rescue ordinary reality David Bohme constructed an interpretation of quantum theory in which the electron is a particle. These particles have at all times a definite position as well as momentum. Also each particle is connected to a special new field that he called a pilot wave. This guides the electrons' movement by a new law of motion. In this model, wave and particle are real. One catch is that the pilot wave is invisible. Maybe we should speculate about the virtual particles as the pilot wave? Bohme states that the wave can be observed indirectly by its effects on the electron. A motion of virtual particles would indeed affect an electron particle. A virtual electron would still have a magnetic moment even though its lifetime is extremely short. Bohme says that quantum stuff has two separate entities. It has a wave and a real particle. He has eliminated the Copenhagen substance that combines wave and particle in a single entity.

Bohme asserts that his pilot wave acts on a precursor, or check of the environment. Environmental changes alter the wave shape whenever changes occur anywhere in the universe. This change in the pilot wave is sensed immediately by the electron anywhere in the universe, and altars its position or momentum or both. If an investigator measures for momentum the pilot wave assumes one shape and if someone measures for position, the wave will be shaped differently.

The electron takes on different attributes for different measurements. According to Bohme, the reason for this is the influence of the pilot wave. For the electron to be classified as an "ordinary object," its attributes must be innate, meaning they must belong to the electron. These attributes are rendered contextual by the pilot wave. In order for the wave to manipulate or inform the electron which attributes to divulge, this signal must become superluminal at times. This faster than light signaling is forbidden by Einstein's special theory of relativity (for space time particles). We are now at a place where Bohme indicates that a quantum particle such as an electron is truly a *thing*. By this I mean a real solid piece of matter. Associated with this particle and all particles is a pilot wave. This wave senses in some way what is about to happen in the vicinity of the particle and signals the quantum what aspects to reveal.

To make things more complicated, those control signals can travel instantaneously across universal space. Since faster than light movement in any manner is forbidden by Einstein's theories, this view of solid quantum particles is not taken seriously by most physicists.

BELL'S THEOREM

John Stewart Bell became curious about Boehm's theory of quantum particles. He found that the theory did as it claimed. The electron and consequently any quantum particle could be assumed to be an ordinary object and still fit into the quantum theory.

Mathematician John Von Newman previously had shown to the satisfaction of most physicists that quantum stuff was not ordinary material. Bell then assumed that the fault lay not in Bohme's proof, but in Von Newman's. One of Von Newman's objections to quanta being ordinary material is that it is unreasonable to believe that an invisible wave could sense the configuration of a measuring device and signal the quantum to adjust what it shows us.

Bell was able to show that an ordinary model of reality as well as a contextual model, has to be non-local (space less). In local reality, it is necessary to state that influences can't travel faster than light. Bell's theorem argues that in any reality of this sort (ordinary matter); information could not travel fast enough to explain the quantum facts. Bell assumed therefore that reality must be non-local. The theorem requires that its objects, quantum stuff, be connected by influences and information that are not local. Bell states that a faster than light communication for ordinary states is a necessity to explain quantum facts.

Bell has proven that the two models of reality, ordinary and contextual must also include signals that travel faster than light. If this is so, our reality consists of more than the four dimensions we are acquainted with.

At this point it is necessary for me to review some of the elemental features of electrons, so that later we can discuss atomic connections with other dimensions. A very simple experiment using an electron gun (like the gun in the picture tube of your TV set), a barrier with a tiny hole in it, and a phosphor screen will give plenty of evidence of the electron's particle nature. With one hole in the barrier and proper adjustment of the brightness, you can see the electrons striking one by one as a particle should. If we were to put another hole in the barrier, (some ware on the same horizontal line), we would begin to see bars of electron hits on the screen. Some of these bars would be of a light shade, and some darker. This interference pattern is what would be expected with a wave output from the gun. This is the same gun we used before but the barrier now has a second hole. If we put a piece of tape over the second hole, the pattern will go back to a particle indication. That electron knows that there is only one hole when it leaves the gun. The Copenhagen interpretation of quantum law says that reality is created by observation. Waves striking two holes should create interference patterns, so it does. Just moments before the same gun was only putting out individual electrons and no waves. If you feel frustrated with this type of weirdness, don't feel bad since Einstein didn't like it either.

Most physicists feel that an electron is a point particle. That is a particle whose real dimensions are zero. It is difficult to visualize a particle with no dimensions at all, but it works mathematically. All of the rest of the elementary particles seem to be point particles also. Atoms however begin to show more definite size. In this case return your thoughts to our discussion of space. A point particle has no size (space). Two points will define a dimension (length), and will indicate space. An atom of Hydrogen (one electron and one proton) will occupy space in our space-time dimension. Electrons by themselves seem to be pure quantum stuff. By that I mean all field, focused to a point. Photons, quarks and all elementary particles are likewise quantum stuff. They seem to be fields of information, radiating from a mathematical point. Some physicists believe that objects like coffee cups, trucks and planets would exhibit quantum wave properties under the right conditions. Limitations of wavelength and instrumentation make it impossible to observe. It seems that elementary particles behave like particles if we watch them, if we aren't watching they behave like waves.

FIELDS OF FORCE

If quantum particles are indeed mathematical points, where does the electrical energy in an electron come from? You will recall from your school days that electrical potential is described as so much positive, or so much negative from some mutual null point. For simplicity, the positive portion of this potential is thought of as a lack of electrons. In our physical world we deal mostly with electrons, protons and neutrons in chemistry and physics. They are described variously as particles, waves and fields. Quantum physics has said that the particle and wave are two aspects of the same phenomena. A field is described as a distribution of force in space. The field of an electron can be measured as electromagnetic. (Think of it also as a field of possibility.) James Maxwell found that if you move or shake a field, it makes waves. In our space-time dimensions these waves move at the speed of light.

Our sensory systems employ electrons to send information to our brain, and electrons to activate muscles in response to that information. During the thought process there are trillions of electrons moving about. An electron itself is a field of waves. As it moves it creates new and different waves. These Quantum waves are all oscillations of possibility. This is where our observation gets tricky. Remember that one way of looking at quantum fields is that they contain all of the possible results of any action. These possible results are available from that other dimension, or those dimensions that our quantum particles inhabit.

Experimentation has shown that the force of our consciousness enhances one possibility from all of the others and makes it a reality in our world. *All* of the possibilities are expressed in the quantum dimensions, and await conscious choices to become *real* in our physical world.

Quantum waves carry no energy, but their amplitude is a measure of possibility. If there are a lot of electrons doing the same thing, they will emit waves that are "in phase." In phase waves tend to add amplitudes. Large amplitudes increase the quantum possibility. If I seem to over stress these points it is only because they will be important when we discuss possibility/probability waves and how they become reality. An interesting point is that the quantum wave function is labeled PSI. It is the twenty third letter of the Greek alphabet. The quantum wave is also very important in individuals' PSY possibilities as we shall see later.

Quantum waves carry no energy while ordinary radio waves (for instance) gain energy as their amplitude increases. They also gain energy as their frequency increases. If you square the amplitude of a quantum wave you simply increase its possibility. The larger the quantum wave, the more *real* it becomes.

An interesting point is that if you have a lot of possibility wave, it begins to change to a probability wave. Since that is so, wouldn't a large probability wave change to a wave of reality? If a lot of conscious entities were exposed to certain sensations and their nervous systems (electrons) began to oscillate (wave) at a similar frequency, would that turn a possible tree into a real physical tree?

Physicists call this possibility wave by different names besides the PSI wave. Einstein liked to call it the ghost field. Some call it the empty field. French scientists call it the presence field. All of these titles have kind of other worldly sounds.

Since quantum possibility waves add when they are in phase, they also decrease when waves are out of phase. This seems like a good reason to get your friends to think in a positive way with you to change a possibility into a probability.

INFORMATION OF
IMMORTALITY

The question of our immortality at present is largely defined in the various religions. Science can indicate with very good authority that the materials of our bodies are ancient, and recycled.

To prove that our consciousness (soul) is as old as the expanding universe, is a bit more difficult. It will be necessary to absorb a lot of new information before we become comfortable with a scientific view of immortality.

Recognize that this information is the result of years of painstaking research by the best of the worlds' physicists. These gentlemen never take anything for granted and never take anything on faith. Not only do experiments have to work out physically many times, but they have to make sense mathematically. Mathematics has been called the language of God by Einstein and others. Maybe that god is really a super consciousness as envisioned by Carl Jung and others. Freud and his followers rejected that theory, and preferred to retain their religious view.

This super consciousness would be the total awareness of all of the entities in the universe. If you envision all of the galaxies and stars in the universe, and all of the possible habitable planets, you can get an idea of the scope of this consciousness.

As you read the various books about the experiments of our scientists, let the information enter your memory so you can bring it forth when needed as a counterpoint during discussions of immortality.

Sleep and
Other States

The main question is why do we find it necessary to sleep? There doesn't seem to be a firm answer as yet. Each of us as individuals can attest to the fact that after a certain period without sleep, every cell in our body cries out for the rest of unconsciousness. Medical science knows that sleep is necessary for our mental and physical health. Without it we become physically exhausted and mentally unbalanced.

What is it that happens to the body when we sleep? Physically we are aware that our eyelids close and our pupils become small. Our glandular secretions fall to a low point. Respiration and heart rate slow down perceptively. Our brain waves (as indicated by an EEG) change character and we become unconscious. This state of unconsciousness is modified by the fact that certain stimuli can cause a return to consciousness quickly. This is not the case when drugs or injury are the cause of the unconscious state.

In sleep, the EEG will indicate many different wave shapes and amplitudes. One wave form is so easily recognizable that it has been given a name. The *Alpha* wave can be intercepted most easily by our instrument at the back of the head, and also at the forehead. This wave varies a little with individuals, but will average about nine to ten cycles per second. It has been learned through experimentation that a wave of this nature indicates relaxation and disassociation. There is a lack of sense stimulation and cortex stimulation. In other words there are no unusual sense impressions (the eyes are probably closed) (not always) and the brain is functioning in an idle condition. We find however that when the eyes are opened this Alpha wave will disappear. This is not true however if the individual is in a deep altered state.

While these processes are underway in the physical body, there are complimentary changes occurring in that part of our consciousness that occupies the quantum dimension. This seems inevitable because of the Electro-magnetic variations impressed on the physical particles as well as on the virtual particles that are in adjacent space and between the atoms of our body. (More about the virtual particles later.) Keep in mind that the electrons that are racing around in our brain conveying information are also a part of those other dimensions.

Visual information stimulates the brain to search its memory files of impressions for similar arrangements of energy levels and thereby recognize (re-cognize) the visual signal. The brain in this case operates very much like a receiver and transmitter. The elemental arrangement of energy patterns is converted by the eye and optic nerve to a coded impulse that soon arrives at a point in the brain reserved completely for this type of signal.

While this process is under way in our physical body, I feel that there are complimentary changes occurring in that part of our consciousness that occupies the quantum dimension. This seems inevitable because of the electro-magnetic variations impressed physically as well as on the virtual particles. Keep in mind that the electrons that are racing around in our brain conveying information are also a part of those other dimensions.

In the conscious state the wave traces on the EEG are complex, and change with sense input and motor output. The change in the character of the EEG trace is merely the electrical result of an increase of brain activity. When the brain is occupied with normal conscious activities, the measured waves are very complex. When we relax or close our eyes the complexity will decrease and the Alpha wave may appear in short bursts. Even with the eyes closed, if the brain is given a problem to solve the wave complexity will increase, and the Alpha will be minimal. Each sensory input requires an action by that section of the cortex responsible for that sense and associated systems. Just as a heart specialist can look at the trace of an EKG and diagnose a heart problem, so can a neurologist diagnose conditions of the brain by viewing an EEG trace.

As we approach sleep the indicated brain waves become longer and slower. In deep sleep the waves will register from one to three cycles per second. The appearance of the waves alone would suggest that the slow cycle of sleep is a natural variation of energy level. There are intermittent bursts of activity that are called spindles. It is my opinion that this is a sampling system of the quantum consciousness. It is perhaps significant that the spindle can be most readily measured at the front portion of the brain. It is from this location that individual traits and emotions are controlled and the Astral body is said to emerge during astral projection.

What causes sleep? We all recognize for instance that monotony or disinterest bring about drowsiness and sleep unless one or more of our senses receive an impression that is stimulating to the brain and therefore to the body. By stimulating I mean an input that necessitates correlation of data in order to initiate the proper action.

I compare the brain to a transceiver in our satellites. The brain receives information about its surroundings in space-time. After comparing new information with old (memory), the brain sends it on via its quantum connection to the dimensions that await its information.

Experiments with animals (cats and rabbits) have shown that if an incision is made between the cerebral cortex and the brain stem, the animal will remain alive but will not react to stimulus. The pupils of the eyes were very small and the EEG trace was indicative of sleep. This is called "An Isolated Fore-Brain." In other experiments, an incision was made at the upper end of the spinal column. This effectively separated the whole brain (Cortex and Brain Stem) from the rest of the body. It proved quite different from the isolated fore-brain, in that the EEG indicated periods of wave patterns that normally are associated with alert consciousness. These periods of wakefulness would alternate with periods whose EEG would indicate sleep. What has been proved by experiments such as this? It proves only that there is some apparent connection between the brain stem and control of our physical consciousness.

Further work by the sleep labs indicated that sensory impulses branch off on their way to the cortex and enter a specific area of the stem called the Reticular formation.

It was necessary to demonstrate that the reticular formation controlled the sleep cycle, and to do this another experiment was performed. For this purpose two groups of animals (cats) were selected. One group had the reticular formation surgically isolated from the sensory stimulus. The impulse to the cortex however was kept intact. In the other group, the impulse to the cortex was severed *after* it branches off to the reticular formation.

As a result, the former group of cats remained in a perpetual sleep with EEG traces that would be expected during sleep. This group had information going to the cortex, but none to the reticular formation.

The second group whose sensory path to the reticular formation was intact, but whose path to the cortex had been severed, would sleep for a time and be awakened by a noise, or they would awaken spontaneously. Although they were evidently unsure of the world (senses removed from the cortex), their sleep mechanism functioned normally. The amount of stimulation received at the reticular formation seems to be the controlling factor of sleep or wakefulness.

Let's put all of this into simpler language and look, at it not with a medical eye, but with the people in mind.

Earlier when we were discussing hypnosis I said that relaxation reduced the amount of information received from your senses. This is accomplished by purposeful relaxation and concentration on inner rather than outer impressions. This has the effect of severing the brain and the brain stem at the top of the spinal column. This is true by degrees as you relax and move more deeply into hypnosis. This increasing isolation (lack of sense impressions) at the brain, will result in a deepening hypnosis or meditative state.

In this state of affairs our consciousness pays more and more attention to memory and other quantum information. This phenomenon has a great deal to do with the accomplishments of hypnotic states. Age regression for therapy and indications of reincarnation are among these phenomena.

There is agreement among the laboratories studying the phenomena that the reason we sleep remains unknown. Medical researchers feel that there must be an as yet undiscovered chemical control in the body that serves as a triggering mechanism. *All* of the functions of the body are chemical and electrical in nature.

The major portions of the body don't need sleep. Periods of rest they do need, but not sleep.

Experimentation has shown that the brain *does* need sleep. For all of these experiments I recommend you read: "The Third World Of The Mind! Sleep", By Julius Segal and Gay Gaer Luce.

Throughout this book we discuss the idea that the brain is an instrument whose job is to maintain the physical portion of the link between the body and our eternal consciousness. Does our consciousness need time to rest? Are the chemical functions becoming depleted or are there other missions that it must perform? Perhaps the body is allowed to idle while these other missions are accomplished. What is the *spindle* that is observed during sleep? Is this burst of activity in the slow wave of sleep an indication that the consciousness is monitoring the body periodically to assure its proper functioning and safety?

Up to this point I have described only one portion of our sleep, that of the slow large amplitude wave function. What is happening when the slow waves are not present? We know that when we are awake the EEG will indicate a multitude of wave shapes and frequencies. In the awake state the alpha waves come at random intervals and in very short bursts.

It appears that some functions of the brain require momentary disassociation from the physical. The alpha wave is generally present when we are relaxed and remembering or dreaming.

If we constantly monitor the EEG from the wakeful period to the slow wave, we will notice that the rapid constant wave of wakefulness will change gradually to irregular little waves.

These are the indications of drowsiness and it is during this period that we experience rather strange phenomena.

We are neither *awake* or *asleep* in this stage of our mental functioning. Our consciousness is accepting information from both worlds of our being. We may "talk" during this period, and some of our words may make sense and some not. Visions may appear that seem quite real. I recall once during a short nap I opened my eyes and saw a big rat on the wall. It startled me and I threw a pillow at it. The pillow hit only a blank wall. The rat was of some other reality.

Our bodies may twitch and jerk in an effort to follow some physical thing that is happening in dreamland. This is an indication that we are so near to the conscious state that when we move or speak in response to a dream that it also registers in our conscious mind. We wonder then how we could have believed we were anywhere but in bed. We alternate between wakefulness and drowsiness until finally our EEG indicates the big full waves of sleep. We have drifted into a natural unconscious state. In this condition, we are more able to become aware of the infinite waves of possibility that reside in the quantum dimensions. These possible events that may not have happened, or may not happen are loosely woven into dreams that keep our brains active during our sleep state.

Dream Time

In adults, experimenters find there seems to be a natural sleep cycle whose periods last roughly one and one half hours. These cycles vary widely but this is a good average.

After a sleeper enters a period of big slow wave sleep, he/she seems to remain there for that period, after that time though the EEG begins to change.

During a period of approximately ten minutes, a sleeper will move up from stage four delta sleep through lighter and lighter stages. Eventually the sleeper will reach the light sleep of stage one. Most of the time this stage one sleep will again be modified into an even lighter stage called *Rapid Eye Movement, or REM* sleep.

This is the stage of sleep during which you have most of the dreams that you remember. It is actually a hypnotic state during which you are very close to wakefulness. Usually if you don't begin to dream during this stage, you will wake up and wonder why you woke. I have shown many mothers and fathers how to take advantage of this state to give their young children suggestions that will modify bad habits and traits.

REM sleep is called that because during this stage the eyes move beneath the lids as if they are following movement. If sleepers are awakened during this phase and questioned, they will invariably say they have been dreaming. Dreams that occur during this stage of sleep seem to be the most vivid or at least the easiest to remember. There is evidence shown by the EEG that dreams do occur during deeper stages of sleep but they are harder to remember. It's almost as though at that time we are farther away from physical reality.

Is it necessary for us to have more than one type of sleep? The sleep laboratories aren't yet ready to tell us, so we will turn to psychology and also the psychic sciences for possible answers.

First of all is it necessary for us to dream? If so, why is it so? Sleep scientists have discovered that an average person spends approximately one and one half to two hours per night in the REM stage of sleep. During this time vivid dreams always occur.

They have also discovered that if sleepers are monitored, and denied this REM sleep, certain mental irregularities begin to occur. If this REM sleep denial is extended, these abnormalities could result in permanent damage.

No one is certain if it is the dreams that are necessary, or just that portion of sleep in which dreams occur more vividly. Is it possible that during this period our consciousness is sort of clearing the circuits in preparation for another day? Is it possible that during this period of sleep we are experiencing memories pulled at random from our unconscious?

I have said many times to my students that not only do a lot of dreams make no sense at all, but words and even sentences spoken during dreams also make very little sense.

Since I have a continuing interest in dreams and consciousness, some time ago I wrote down one example. My alarm clock went off one morning and apparently roused me from a REM period of sleep. A dream and a portion of conversation were very fresh in my mind, so I wrote it down for later study. What I wrote was—"gee Jeffrey did nothing but a boob suspicious thing." The only thing that makes any sense to me at all is the name "Jeffrey." I have a son by that name. As far as the rest of it, your guess is as good as mine.

If scientists have determined that REM sleep is necessary, how about the other sleep conditions? Most books have a singular lack of comment on the need for slow wave sleep. Apparently we do need this type as a precursor to the REM sleep. I mentioned before that when the EEG registers the slow wave sleep, our heart beat is regular and some of the muscles of the body (throat in particular) are tense. These and other body states tend to show that during this sleep period our consciousness is exerting the least amount of control.

It is during this type of sleep that *out of body experiences* take place. It is at this time when our psychic body is elsewhere that it is only connected to the physical body by that silver cord that astral travelers speak of. (Read—"Psychical Phenomena And The Physical World" by Charles McCreery). Are the spindle bursts that we measure in the EEG our consciousness checking up on its physical counterpart? Isn't it convenient that this period of sleep leaves us with very little memory of such strange events?

Drowsiness and REM sleep seem to be very much alike. During the period between drowsiness and sleep a high incidence of hallucination occurs. These two periods seem to be the times when our consciousness is present, but is exercising a minimum amount of control over certain centers of the brain that comprise the unconscious.

To those of you who have had occasion to observe a sleep-walker it will come as no surprise that they aren't really *with us* during those wanderings. Their mental state is akin to a medium stage of hypnosis, where some control of the body remains, however the mind is creating a world of its own. It is significant that sleep-walkers as a rule do not remember their escapades. This is not always the case, but most do not remember talking or moving. If we equate somnambulism to hypnosis, it is well to remember that only in the deepest states of hypnosis is an automatic memory loss observed. Not surprisingly somnambulism is derived directly from deep delta, or stage four sleep.

The retrieval of memory requires that we disassociate from our physical sensations for a moment. This enables our brain (quantum particles) to search for the information we require.

If an individual is monitored with an EEG, and asked questions needing answers from long term memory, you would notice short bursts of alpha waves on the recorder. These waves occur during relaxation, sleep and memory search.

Consciousness is awareness. We can be aware of physical things, non physical things, or both at the same time. We can go back in time to remember our youth and we can attempt to visualize our future. Our past is hazy with intervening memories and our visions of the future are cloudy with many possibilities. The past is now a part of our physical space-time universe, and the future is still all quantum possibility and choice.

MATTER/QUANTUM DUALITY

Bell's theorem states that all quantum entities are in constant communication with all particles in the universe instantly.

If every *particle* in the universe is related in that intimate way it sounds very much like the association in an organic unit. Quantum mechanics seems to say that all things (including you and me) are a part of one pattern. Paul Davies and John Gribbin in "Matter Myth" say that since macroscopic objects have associated waves, "the independent reality of everything seems to go into the quantum melting pot." Always remember that the phase of quantum waves is of paramount importance. Waves that are *in phase* become more powerful, and waves that are *out of phase* are less able to manifest in our material world.

The electron particle that is so important in chemical combinations is really an area of peak wave amplitude. These particles can only appear to occupy certain energy levels in an atom. This is where the vision we have of electron orbits comes from. In reality the waves (of probability) interfere with one another at most areas around the atom's core, but at certain increments the waves are in phase and add and manifest as particles.

This phase amplification may have been one of the experiments you did in your high school or college physics courses. It is a rather easy test for light interference using the quantum unit of the photon. If you have ever seen these tests, or done them, you will remember that the experiments cause light bands to appear on the receiving screen in a specific sequence. There will be a band with a lot of light, then several that are dimmer. This pattern will repeat itself on the receiving screen. These bands with lots of light photons are analogous to the energy levels of the electrons. Where the waves interfere the particle aspect of the electron does not appear. In areas where the waves (of possibility) add, their amplitude is great enough to manifest as particles. They really aren't anything solid, but our senses and methods of testing say they are.

This adding and subtracting of phase relationships is the reason for things happening or not happening. The possibility of an event occurring is dependent on the amplitudes of the probability waves.

As I have mentioned before, our process of thought causes a great deal of activity in the movement of electrons. This is true not only in the area of the brain, but in the whole body. The motions of these electrons have an effect on virtual particles as well and the combination will modify your local *field* as well as other places and times. This doesn't happen in isolation, however. Any conscious entity will also tend to modify the wave shapes locally as well as far away simply by *their* thought processes. The combination of all conscious entities associated with a problem, a question or some deed will decide what possibility will be enhanced to a probability.

I have mentioned on other pages that consciousness can be equated with awareness. Awareness extends very far down the scale of animal, vegetable and even mineral. A blade of grass is aware in a minimal way of light, heat, water, etc. This awareness is seen as chemical activity. Minerals react physically and chemically to heat and cold. Some even change their polarity in reaction to a strong nearby field. Single celled plants and animals have awareness as well. They will move their cilia in an effort to approach and capture food. Crystallization is not a thought process but the element involved has some kind of awareness of conditions, and will begin arranging its molecules as is demanded by its chemical consistency. Chemical activity is advanced or retarded by the movement of electrons.

It may be a help for us to think of quantum particles as tiny worm holes in space. Or we can visualize them as naked singularities of quantum size. Our space can be influenced, or can influence other dimensions through these space-time tunnels.

Quantum mechanics sees the physical world not as *things*, but as relationships of waves or fields influenced by the force of awareness, or consciousness if you prefer.

Think of a flower in a field. The sun rises in the morning and the flower slowly turns to face the sun. This can be explained chemically, but we have to go farther. The directions it could have turned were all possibilities until certain stimuli arrived. These photons and the energy of warmth modified the waves of the chemicals involved and caused the shift in direction.

Is instinct somehow encoded in our genes, or is it a, group memory available to us through time and space via our connection with the infinite information of the quantum dimensions?

John Bell tells us that each particle is part of the whole of time and space and is instantly in communication with the infinite. There is a discrepancy in the way we look at the world. We are educated to believe that something is there, or it is not. Our sensual experience tells us that the world is solid, substantial and *out there*. Quantum mechanics says this isn't so.

The rule seems to be that as the quantum wave becomes greater in amplitude, we begin to bump into that information. What I am saying is that directed consciousness, whether animals and grass and trees or you and I are what determine the firmness of the fields (ideas) around us.

The universe around us appears to be made of *stuff* that we perceive as good solid touchable matter. At the same time the scientists we trust to tell us the truth, say that this isn't so. The world isn't what it seems. Their conclusion isn't just that maybe there is no solid matter; they say that there definitely is no solid matter. This statement can be made with confidence since the particle/wave duality is the characteristic of everything, as Gary Zukav says in "The Dancing Wu Li Masters."

Quantum Waves and Virtual Particles

Serious scientists as a group gradually discarded the concept of an Ether some time after Michelson and Morley's experiment and Einstein's theories of relativity.

Quantum mechanics however maintains that all quantum particles are truly waves of possibility. To visualize a wave of possibility requires a big stretch of the intellect. The word *wave* is usually used any time that a cyclic rhythm in a positive and negative direction occurs. We are all aware of the 60 cycle current in our homes, and this is simply sixty 360 degree changes per second. The electrical potential changes continuously in a sine wave pattern around some null point, or around ground potential.

Transpose this mental picture to a wave of possibility. What this says is that there is a specific range of possibilities and the electron (for example) cycles through all of them at some specific rate.

It is very difficult to visualize possibilities cycling in that manner. It is not as difficult to visualize a *force* cycling. In our space-time universe we are aware of the vibrations of electromagnetic forces. These forces are the basis of matter and time. Photons have various rates of oscillation, and are identified by their frequency. An atom of any element has a specific rate of oscillation, and the frequency of each element is different. This is the basis for our photo-spectrometers. The electron is the *particle* we associate with the force of electromagnetism. Physicists assure us that an electron isn't a thing, but is in reality a point of quantum possibility.

Electromagnetic force (the electron) is clothed in two suits, both *force* and *possibility*.

Virtual particles (ghostly electrons, photons and all of the so-called particles) are exactly the same as their space-time counterparts except they appear and disappear so quickly. Is it possible that those little worm holes to another dimension (all of the quantum particles) both regular and virtual are the medium of possibility?

An electromagnetic wave of an electron could conceivably pass among, or through a field of possibility and by its potential (strength or amplitude) cause a possibility to become a probability. A photon or virtual photon could do the same for a time interval.

In an atom of anything, the so called orbits of electrons are regulated by the quantum waves. In specific areas out from the atoms core the waves of possibility interfere with one another. This is possibly because of the reaction with virtual particles. At some precise point however the waves add rather than cancel, and a maximum possibility wave exists. This strong possibility is seen by us to be a particle at a certain distance from the proton (in the hydrogen atom).

The Schrodinger wave equation provides a consistent size for the hydrogen atom. One electron and one proton at the ground state (energy level) predict the size of the perceived orbit or wave maximum.

Detailed work by the quantum laboratories over the past forty years indicates that quantum entities in their natural state are wave like. They exist as a composite of all they can be. This view is compatible with the possibility of the electromagnetic wave moving among the infinite possibilities of the quantum dimensions. If we carry that through to its conclusion it follows that the physical world we inhabit is a tiny portion of its potential. Experimental physicists tell us that their lab tests indicate their preparations and thoughts dictate the results of the tests. These conclusions have been reached through unnumbered tests in the laboratories of the world. I believe that this is the result of an interference wave set in motion by brain activity. In the past I have compared electrons and virtual particles to little space-time worm holes. Sometimes it helps me to think of all quantum entities as mathematical points that broadcast a wide spectrum of possibilities. When a conscious entity is present, their awareness results in waves that interfere and cause patterns of addition and subtraction. The resultant wave peaks will be what we perceive as our physical world. Remember once again that this conscious agent need not be an "intelligence" as we conceive them. Any sensate plant, animal or whatever will cause the same thing. Particles are the result of wave interaction.

The strengthened field that results from billions of strong wave crests is the matter that we sense. The more electrons involved, the denser the field. Hydrogen gas is composed of one electron and one proton. Billions of them in combination still produce a very tenuous wave that we move through easily.

Hydrogen when combined with Oxygen in the proper fashion produces a wave of greater amplitude (water), and resists more strongly our movement through it. The analogy can be continued to the atoms we consider solid *stuff*. As we move up the periodic table, the numbers of electrons, protons and neutrons increase and at the same time the degree of *matter like* increases. This is an obvious result, and can help in visualizing the change from no physical matter to a very resistive matter.

As you read it is good to remember that I am simplifying quantum interactions greatly. However, the end result without considering spin, photon emission and so forth is what is interesting.

A short time ago I was discussing altered states of consciousness. I hope I was able to reassure you that altered states are normal functions and are easy to achieve. There is a lot of good evidence that consciousness is a quantum process. This being so, it is also a multidimensional state.

Many of the therapy sessions that I conduct last an hour or more. At the end of the session it is common for the patient to remark that it seemed like only minutes. Unquestionably there is time distortion during the altered state, just as there is during the unconscious state.

During these altered states, if given the proper information and suggestion, the individuals can begin to control many physical systems that previously were the domain of the unconscious. These include blood pressure, heart rate, respiration, pain and so forth. The change is accomplished by a modification of the thought process. Thoughts remember are movements of electrons in various areas of the brain.

Movement of electrons causes various distortions in the quantum field. I really don't like the word *field* in this case because it tends to make you think of physical things. The electron movement disturbs the non-physical quantum waves of possibility. This disturbance is aided by the virtual particles that are present. Since virtual particles are really not considered part of our physical universe, they need not follow our physical laws. Continue to keep in mind that the work of John Bell and many other prominent physicists tells us that all quantum particles are in intimate contact instantly throughout the universe.

A conscious thought makes changes in the quantum wave. It does so because electrons tend to repel from one another when they get close. The physical electrons (permanent in our space) move through areas that are also occupied by non-physical electrons (virtual particles), and this warps the quantum possibilities.

For a long time, analytical physicists believed that virtual particles would not interact in any way with physical systems.

Most are aware now that this isn't quite true. There are certain very brief interactions between some physical entities and virtual particles.

The important reaction isn't physical at all. The disturbance to the quantum field is the magic part.

This quantum field is all of the information, experience and wisdom of the universe. This information may be a super consciousness, or God, or something else altogether. The process of making choices and thereby causing a flux in this field of another dimension can possibly make a reality out of a possibility.

Some physicists that prefer to think that we live in one unit of an infinite, number of universes (part four of the Copenhagen interpretation) would say that any choice we make creates an entirely new universe.

Quantum physics calls the particles we are made of *waves of possibility*. Another way to think of them is a dimension of information. Your body and mine are expressions of just a tiny bit of that information.

In the June 1992 Omni magazine, there was an article by Bill Lawren. It is about a French immunologist that feels he has uncovered the mystery of "molecular memory."

Dr. Jacques Benveniste, who directs the French National Institute of Health and Medical Research, has voiced an opinion that runs counter to present medical feeling. In his research he feels that there is a "magnetic language" that allows one molecule to record the "essence" of a second. His peer group feels that he has somehow gone astray.

Dr. Benveniste was working with serum effectiveness and felt that modern medicine has been over medicating. He diluted the serum with water gradually up to one tenth of the original strength of the serum. Working with what was now in his words "plain water," he found that this water still acted as if it was the serum.

His peers and other skeptics of course had the usual narrow reaction. They claimed that the experiment must have been done incorrectly or dishonestly. My advice to Dr. Benveniste is to keep up the good work. Become better acquainted with quantum physics and the communication that occurs between electrons. All of the chemical elements he was working with were *imprinted* with the original serum's *probability*. This quantum information survives the dilution sufficiently to act as the original full strength serum.

Information is the basis of our universe. The things we sense as physical and those things we believe are *not* physical are all information. Physicists describe the quantum wave/field, as a wave of information. The desk I write on is the result of a consciousness in the past imprinting a lasting memory on the elements it is composed of.

THE PSI FACTOR

Many of the books about the connection between quantum physics and philosophy discuss the question of *reality*. They ask questions such as, "are things really there if we don't look at them?" After years and years of research and thought, my answer is *yes they are*.

In our text so far I have discussed briefly that the waves of possibility that are produced by a consciousness (everything that can react in some way to information) react with the quantum's infinite information to produce particles or waves with a well-defined memory. The cabinet maker had something very definite in mind as he/she constructed this desk of mine. These conscious ideas remain with the wood as long as the elements retain their integrity. Just as a marble statue captures the thought of the artist, the trees in your yard, the pan on your stove and the road to work, all have been sensed in many ways by many types of consciousness. All of these impressions result in a very strong possibility wave that is maintained by the "molecular Memory." Many times you will hear people say, "I have a strong impression that this is so." Mountains and trees just happen to be extremely strong impressions. Keep in mind as you continue to read that reality and matter are the result of conscious entities increasing the amplitude of possibility waves. This process is relatively the same as amplification is in our physical world. In phase waves add and increase amplitude. If a lot of conscious entities think in the same way, the waves will be in phase and the possible will become reality.

Individuals, who because of their backgrounds believe only in physical things, have difficulty with the concept of consciousness. Physiologists and neurologists whose training is centered mostly on the physical for instance explain personality, memory, love, sorrow and all traits of individuals merely as electro-chemical activity that ceases at death. For these people, we are all accidental and useless chemical reactions. They seem to ignore the fact the electrical and chemical functions are all quantum activities and exist also in other dimensions at the same time. I have a feeling that deep down they hope they are wrong. I am sure that many of them hold on to their religions for that very reason.

On these pages I have used the word consciousness many times. It is, a word that can be replaced with soul or spirit or awareness as you read. The word's soul and spirit are generally used in religious context, and are probably inappropriate in a scientific discussion.

All of the various religions of the world claim to know what must be done to keep your soul on the right path.

Even here though there is much disagreement, there is a very large segment of the population that feel unsatisfied with explanations like "it's a mystery," and "you must take it on faith."

One of the brains highest functions are to scan for changes in sense information. This function never ceases, even when we sleep. One reason we dream at night is so the brain can continue to solve problems. An unsolved problem of any kind is an irritant to our brain/mind. The separation of the brain and mind is common in religions and societies that believe in a soul or spirit. It is uncommon for groups and individuals that deny any non-physical segment of life. As you have noted in the pages of this book, *life* is a very complicated concept. Humanity and perhaps some other large brained animals are at a point in evolution where we seek answers to those unsolved questions that irritate our consciousness.

This unsolved question has been with us long enough. In the past and at present the attempt to impose one belief or another has resulted in war and death. It always comes as a surprise to me that to some forms of *"thinking"* is a dirty word; to question a dogmatic practice is *unthinkable*.

Our awareness of the world is always changing. Not to long ago there were particles and waves. Now there are possibilities that react as waves and particles. There are no longer *things* but strongly defined *waves*. These waves are only partly of this world. Their origin is of a place where everything is possible. Time is reversible and choice is infinite. Research may make additional changes.

A quantum entity such as you and I have extensions in both space-time and those quantum dimensions. We are presently aware of the state we call *physical*. We are much less aware of the state we should call universal.

Carl Jungs feelings and opinions of the true state of affairs in his writings and philosophy were widely read. He became aware of the non-physical connections between people, groups and nations. Modern understanding of the interconnectedness of *particles*, redefines our understanding and comprehension of things once called "extra sensory."

In my opinion our senses are universal. Necessity in the past has confined our awareness to our immediate physical surroundings. As we evolve to a more peaceful race, unused areas of our brain will begin to listen more closely to those *whispers of the universe*. Neurologists tell us we use at most ten percent of our brain potential. It is also true that nature never evolves things that aren't useful in some way. That large brain is an example. The great rush of new information and inventiveness will make necessary humanities use of the rest of our heritage.

All we observe in the universe is made of quantum particles. These particles are also part of the unseen world beyond. This un-sensed world that is an intimate part of our body and soul will soon be accepted and studied.

As I write about the quantum and its field of infinite possibilities I become even more aware that we all define our world. Our hopes and our dreams are real in our larger selves. If we believe enough and get our friends, families and nations to believe with us, all things are possible. Synchronized thought will build a wave that will change from possibility to probability.

Remember that the future is not determined, only the past is. We determined the past by choices. By our acts of choice we selected from all of the possibilities, and made one *real* history.

Similar acts of choice will *realize* our future. Let's begin now to listen to those whispers of the universe. All things are possible in the infinite quantum world.

It is very interesting to me that in quantum physics the wave function is generally labeled PSI. Strangely enough, those individuals who through practice in altered states, or gifted through genetics that seem to sense more than others, call this gift the PSY function.

By using the techniques of progressive relaxation, meditation and biofeedback, many wonderful things can be caused to happen. These things occur because we purposely manipulate the quantum system. Remember too that the possibility wave is pure information.

In that realm are the possibilities of marvelous inventions; possibly one that could give us an infinite source of clean energy, or possibly one that could show us ways to manipulate gravity and use it for propulsion.

You recall that Einstein said that gravity is really a warp in space-time. It isn't just a function of space-time like a field, but a manipulation (warp) of space-time. In this case a warp means a bend, or a twist or a curve. Physicists have come to believe that consciousness creates our space-time. Doesn't it seem logical that the power to manipulate our material universe really resides in individual and combined thought? Light and all of the various frequencies of the photon are also a part of space-time. We calibrate time by the velocity of the photon. In reality, the photon does not exist in time. As a consequence of this, space does not exist either. (All individual quantum points are space less and timeless.) If we were to enter the event horizon of a black hole, time would no longer be a consideration. The same would be true if it were possible to hitch a ride on a photon.

Keep in mind that if we stand in our universe and measure the travel of a photon, time is involved. The photon is a quantum particle however, and as such is not a part of this universe except as a possibility. For that reason, space-time does not exist, just as for us the fifth dimension does not exist.

John Bell and others have experimentally shown that there is instant communication from one end of the universe to the other. Superluminal is the word that is used to describe faster than light communication. Since we measure time by light speed, does that mean we can see into the past? When you use your mind to remember something, are you seeing into the past? In our normal conscious state, memory is often cloudy and incomplete. It has been my experience with altered contemplative states that memory is greatly improved. We can remember things we didn't know that we sensed at all. This technique has been used by police at times to aid victims to remember information that wasn't available normally. There is still the question of reliability of the information. By allowing our mind to wander, we can go back very vividly in time. We can experience things that occurred hundreds of years ago. Are these our memories, or are they a part of the universal memory that we have been discussing?

Notice I said we use our *mind* to do this. I am indicating that this is a function that is in part separate from our brain. This is a condition similar to radiation and radiator. Sensual experience causes brain function. This function causes a great number of electrons to flow throughout the quantum fields of the brain. The movement of electrons in the brain transfers information from our space, to the quantum dimensions. This information then becomes a permanent part of our consciousness as well as the total consciousness of the other dimensions.

The brain radiates information via the quantum waves. Keep in mind however that we are speaking about non-physical, timeless and space less radiation.

Because of the laws that quantum physics seems to follow, the phenomenon of ESP is probable. Anything that has occurred in the past leaves an impression in the fabric of space-time (photon radiation mostly) and also in the total consciousness of the universe. That is the reason that trees, mountains and planets persist even in the unlikely event that nothing is observing them. The universal communication between quantum entities allows us to tap into these past events.

Viewing things of the future would require a different set of rules. We (most of us) like to believe that the future isn't set. Many still believe in a judgment day that is inevitable, and the destiny of the soul. For me it is comfortable to believe that the future is the result of our choices. Cause and effect are the scientific terminology. In the quantum pool of infinite possibility, we (collectively) choose to emphasize certain aspects that become our reality. It is my belief that individuals who view the future, do so by evaluating the trends of society and report on the possibilities that may be chosen. It would be a good reason for the hits and misses we see in prediction.

As a therapist using hypnosis I have of course used the technique of *age regression*. Generally it is used to elicit memories of periods when traumatic things occurred that are affecting the individual at present.

Now and then a patient will automatically report on events that occurred many years before their own birth. These *memories* may or may not be evidence of a previous existence. Although I believe that our essential being reincarnates from time to time, these memories are far from proof. We must remember that all space-time things and experiences remain in the quantum field as *things that were*. This total infinite field of information is often tapped into by sensitive individuals. An altered state of consciousness that we use during hypnotic regression greatly increases our sensitivity. These quantum memories of the past may represent an emotional similarity to a present situation.

Some years ago a friend and business partner induced a good deep altered state in me, and asked me to venture into the future to ascertain what my next incarnation would be. Dutifully my mind was flooded with visions of Asian type architecture and people, such as you would see in a news program from Japan. I sensed that my body type was also Asian, smaller boned and more delicate than my own large Norwegian frame. Images of advanced laboratories and computers convinced me that I worked somewhere in a scientific field. As I have said before, this may or may not be an indication of future events. If past, present and future all exist as a unit, then my vision could be true. If the future is already set though, it makes us just actors on the stage of the universe. We will be acting out what must be without any real choice. This scenario doesn't sit well with me, and most of my friends. If the future is yet to be decided by our choices and activities, my vision can only be one of the possibilities from the quantum field.

QUESTIONS

Niels Bohr felt that, quantum physics is sure to be accepted as a part of biological science at some time. I might add that it is already fast becoming a part of our understanding of philosophy. For more information on this, read some of the good books such as; "Physical Cosmology and Philosophy" edited by John Leslie or "The Dancing Wu Li Masters" by Gary Masters.

At this particular time in the history of the United States, science finds a difficult road. Young minds that are so necessary in the search for novel ideas are discouraged by the need to study instead of play. We have a great need for talented engineers and scientists of every kind. My hope is that curiosity about life, death and reality will once again start a drive for knowledge in our young students.

You have been reading about the strangeness of *quantum stuff*. This lack of solidity applies not only to electrons, but also to the other so called particles of our dimension. It seems that matter, like light is both wave and particle. The resemblance between light and matter is greater than we supposed.

Einstein's law of photo-electric effect required that light be composed of energy corpuscles that vary inversely with the wave length. Both wave and particle were once again involved. The same is true of the effect we call matter.

Louis De Broglie felt that if the electron were to be considered a geometrical point, the energy surrounding it would be infinite. He considered this to be an impossible situation. I really don't see why. A geometric point would be the origin of infinite radii. A wave expanding along these radii will degenerate as the square of the distance. The quantum geometric point represents infinity of possibility. Consequently this infinity can be any conceivable space-time value.

Physicists should begin worrying a little less about their equations that end in infinity answers. Remember that the quantum dimension *is* one of infinity. As these equations are worked through there are intermediate points in the computations that give clues to material goodies that are a part of the infinite possibilities.

Remember that the quantum is infinite information. Don't just look at the infinite answer, pay attention to the finite portions of the equation.

There seems to be a kind of fear of the concept of infinity. When we speak of time and space, the word infinite brings visions of unimaginably long and immensely big. The truth is that if I draw a line _____ that line can be divided an infinite number of times. Infinity can be any size, just as an hour can be divided an infinite number of times also. Often in these pages I have said that a quantum constitutes an infinite amount of possibilities, or if you prefer an infinite amount of information. One second after the Big Bang the universe was infinite in size and time.

If we overlook the period that cosmologists call "inflation," space-time can be considered radiation. The size of the universe and the time it has existed are calculated by the velocity of radiation.

If we use the velocity of radiation (photons), every second the radius of the universe increases by approximately 186,000 miles. Cosmologists in their writings make a point of saying that there is no center to the universe, but three dimensional geometry would show you differently. These same scientists also like to say that the exploding space-time universe didn't expand into anything, because our universe is all there is. For some reason these learned gentlemen want to ignore the evidence of quantum science that shows there are dimensions external to our space-time.

It is really difficult to visualize an explosion that doesn't radiate, in all directions. We are asked to visualize a balloon with spots on it. As it inflates, all of the spots move away from one another. But this doesn't go far enough. Our instructors also tell us that the farther away the spots (galaxies) are the faster they are moving (also true).

If the big bangs reaction is creating space, wouldn't you assume that space would conform to the mathematical perfection of a sphere? Even cosmologists keep telling us that at a certain time the universe was the size of a grapefruit, and shortly expanded to the size of a basketball. They are visualizing something radiating from a center. If indeed the universe is spherical, there is a lot of matter 180 degrees away from us that is not observable.

One of the rules forwarded by the cosmologists is that at any given time the size of the universe is judged by how far light could travel since the Big Bang. Once again are we speaking about a sphere? If we visualize time and space beginning at a point and exploding outward, the picture we see will necessarily be an expanding globe. This would be true even from the inside (within space-time). The radiation would go in all directions. If you choose any of an infinite number of points on the periphery of the expanding universe, there is a corresponding point 180 degrees away on the other side of this expanding globe. Since both of these points are moving away from the center point (physics and math will say there is one), is the universe twice the size we have computed? It is necessary that we continue to visualize in three dimensions. An imaginary balloon with spots on it can be close too accurate if we inflate it not from one end, but from the very center of the sphere. If all of the matter in our universe precipitated out of the radiation at one very specific period of time and none since then, that does mean that there is a three dimensional band (the skin of the universe) expanding outward. This would make necessary another condition. After the radiation changed to what we consider matter, an area of empty space is following and expanding along with the material.

It is necessary for you to visualize yourself at the very point of the Big Bang. When you do this it will be easier to understand that the matter is expanding outward (along with space-time), and that the energy of the explosion is enough to create only a certain amount of matter. Space and time however continue to be injected into the expanding balloon of the universe. When physicists talk about space they sometimes remind us that it really isn't empty, but is filled with virtual particles. These are the particles that oscillate between our universe and the quantum realms. This oscillation is so rapid that they can't be considered a part of our space-time. Experiments in our nations labs have shown however, that they do effect our material world in certain ways.

The picture we have been drawing in our mind is somewhat different from the expanding balloon. We do have an expanding ball (space-time), with an outside layer of material composed of galaxies, dust and gas. This expansion being fueled by more and more virtual particles from the quantum dimensions of possibility.

Astrophysicists are limited, regardless of the ability of their machines by the visual horizon. This is the point beyond which we can't see because light has not had time to reach us yet from those distant places. My knowledge of cosmology is sketchy I'll admit, but all of the computations I have read base the size of the universe on the speed of radiation from the Big Bang. Does this measurement include only one radius? How about the diameter of the sphere?

The "Anthropic Principle" says that the age of the universe must be sufficient to allow all of the processes for mankind, or at least a questioning consciousness to evolve. It goes on to state that if the universe were much smaller, or larger, we wouldn't be here to observe it.

Since these rules are talking space-time, it does fit that the calculation from the beginning (sciences Big Bang) to now *is* the interval for every spot in the universe. A point on the other side of the mathematical center would still be the same age. It has occurred to me in the past and maybe you now that in order to get to that center that I have been talking about would require passage through billions of years of non-material space-time. This area would be populated only by virtual particles of possibility.

Cosmologists tell us that each galaxy is retreating at different speeds from one another. This means that there is an angular difference in their travel. Also that this angular difference makes some of these galaxies farther away than others. If you follow this backward in time, you would expect that the materials of the universe are radiating from an original center point. Because I believe that nature prefers the most economical way to distribute energy (the path of least resistance), it just doesn't make sense to me to think of our universe as saddle shaped or fan shaped. By eliminating these other possibilities, that leaves the mathematically perfect sphere shape. Time, in this case, can be calculated from the center of the globe outward on one radius. Probable mass however would be derived from the volume of the globe since energy and finally matter would spread in all directions from the original explosion. When you contemplate the space and matter of our universe, always keep your mental picture inside of space-time, not external to it. This being the case, space and time expand in all directions, and carry the material of the universe with it.

One of the popular theories of our universe is that the expansion will continue until gravity overcomes the energy of expansion, and a contraction commences. This can occur readily with a globular-shaped universe.

Our Dual Body

In order for you to begin to change your attitude about yourself and your place in the scheme of things, your body is really a good place to start. You have to recognize that we (our consciousness) are not native to this space-time dimension.

Your *body* is composed of quantum points that take up *no space* in this universe. All of the leading physicists agree that quantum points are not things. These points collectively *do* outline or indicate the position of your consciousness, and the area of your control. The fields of the quantum stuff are phased (maximized) so that your form in the space-time dimension repels other material things, because the fields are different.

It might help to picture a thin rubber glove immersed in water. Your hand in the glove is the real you, your spirit or consciousness. The thin membrane of the glove can correspond to the quantum stuff that your body is composed of. The water it is immersed in is the space-time universe.

If you can visualize this you can see that your immortal self is separated from space-time by a thin layer of quantum material. The life force remains in the quantum dimensions, but is the director of your body. In our little visualization you can see that if the finger moves in the water (space-time), it is the hand in the glove (consciousness) that moves it.

I realize that we are thinking in a very simplified way, but it is necessary as a beginning for your change of attitude. Recognize that this membrane (the glove or quantum stuff) is throughout your body, not a simple thing like a sheet of rubber, but between each particle of quantum stuff and the space-time dimension.

When this glove (your body) begins to deteriorate, your consciousness pulls back slowly (in most cases) until the shell is left without any driving force. The deterioration of the physical body is caused by other forces that the consciousness is unable to overcome.

I'm sure you have heard many times when people speak of a dead body that it is only a shell, and that the soul has departed. That is a true statement if we are referring to chemical and electrical tests done on the body. The migration of the soul is a subject that is more difficult to prove. I believe, as do many that the consciousness lives on, perhaps to take on a new body at some time to experience new things in a different world.

The quantum particles we have been speaking of are points of a field of information. Maybe we could compare it to a field of radiation where the photons carry the information. At any rate these quantum points are the access tunnels to all of the intelligence of the cosmos. This includes every dimension outside of our own. Remember that intelligence isn't a physical thing and doesn't follow the rules of time and space.

When our consciousness makes a decision, even a little one, that decision is impressed on all of the quantum ports (our body), and the infinite possibilities of that other dimension are amplified by fazing the waves and are made real.

Physicists are experimentally aware that quantum stuff is moldable by our thought process. Nature is very stingy however, and will allow only one bit of information at a time to become reality. No matter how sophisticated the question (test) is, you will get only one answer of reality and everything else is hazy and indistinct.

The exclusion principal is the basis for the rule that says an electron, or any quantum entity will reveal an infinite range of possibility, but can be persuaded to reveal only one *reality*. Since quantum stuff seems to react to consciousness, is it possible that the universe is constructed by consciousness? Are we constructs of a universal consciousness? Do we participate in this construction?

The question is open to you, but from my point of view we really do. Decisions I make effect others. If I decide to play golf on Saturday and the rest of the family wants to go to the mall, there will be a discussion, and then we will all go our own ways. Their reactions to this situation will affect still others. These actions and reactions modify our microcosm. The waves radiate outward shaping our present and influencing the future possibility.

If you believe the teachings of some of our religions, our planet is alone as a home for consciousness. The various religions like to emphasize that point. The idea of course is ridiculous as we become more aware of the probability of many other habitable planets; but that is how it stands right now. Since the planet itself has only been here for approximately 4.6 billion years, what sensate system organized the other ten or eleven billion years of universal expansion? Science has found that there is almost no room for changes in the progression of the universe or we would not be here to observe it. Since many in science have decided that matter does not exist unless it has been consciously produced, who or what did that?

One of the questions prompting this book was, are we immortal or are we not? I indicated that the particles of your *physical* body are indeed immortal, and have gone through many incarnations as particles. On various pages I have paraphrased many noted physicists. They state that once an electron or any quantum has been in contact or proximity to another particle, they are forever in communication. This connection can span the universe. The quantum particles that constitute your body originated in the birth of the universe. The heavier particles were given birth at the center of some huge star that spit them out in a super nova explosion.

All of these quantum particles have an enormous history of contact with other particles. Each contact increased the effective range of the particles' communication. If we confine these possibilities to just our planet, we are still an extended part of a great deal of that planet. For instance the water of your body has been the same water in numberless places.

The same is true of all of the atoms of your body. If the science of the quantum is valid, our bodies (physical and non-physical) extend throughout earth and space.

Our self awareness is a consequence of the cohesiveness of the quantum stuff we are composed of. Remember that this quantum stuff appears solid, but is not. Your body is instead a field of waves. These waves of possibility exist not in time or space, but merely influence those dimensions. Your body and all matter seem solid because quantum points are exclusive. These entities tend to push away other points.

Dissolution of the physical body occurs in the presence of strong possibility waves whose phase runs counter to your present manifestation. This is a product of the interference mechanism we have talked about before.

Individuals who practice the art of acupuncture tell me that these points are centers of radiation; small areas where our lives force are at a maximum. The needles or pressure (of acupressure) stimulate activity of the field and aid in healing by rebalancing the field of radiation.

Some equipment does exist that can analyze the fields of the body. The EEG and the magnetometer are two examples. These physical fields are the result of movement of electrons. Not only are their electrical changes, but a consequence of this is a magnetic flux.

If inspection of the body field were possible (it probably is) it would probably look chaotic. The acupuncture points we mentioned could be those points in the chaos that add together and peak. It is possible that an injury or illness disrupts the normal flow of the field and causes distress.

A conscious entity must have an integrated field of quantum forces in order to manifest as a physical body in space-time. Disruption of this carefully constructed field makes it increasingly difficult for our other dimensional entity to communicate in a physical sense.

Our physical body and our non-physical body are one and the same. This statement isn't to be taken on *faith* but to be researched by scientists.

Physical scientists that are current with the investigation into quantum facts, all agree that matter is not what it seems. Its apparent solidity is a function of wave phase and amplitude. The phasing's and resulting amplitude seem to be the result of conscious shaping. The process of thought forms a matrix of waves that constitute our physical body.

I am not bypassing the chemical processes of the genes and elements. Keep in mind that these things are all quantum stuff, and are under the influence of long lines of predecessors as well as the pressures of conscious manipulation.

The Brain and Evolution

Let's consider the human race as a whole for a moment. The evolution of our species is no longer a theory, but a fact in scientific circles. Remember that a theory is only that until it has been sufficiently tested to insure that it is indeed a fact. The *theory* of evolution has been tested by the world's foremost scientists through most of this century, and has stood the test. If you would like to check my statement, please read; "Science And Creationism", edited by Ashley Montagu. This book was written by some of the best of our scientists. Einstein's equations of the equivalence of energy and matter are still called, a *theory* but I beg to differ.

Purposeful evolution requires a need, or an advantage in some form. Throughout these pages I have spoken of consciousness basically as awareness. It is possible for an electron on one side of the universe to be aware of the condition of another electron on the other side of the universe. Awareness is universal and is not subject to space-time rules.

Early hominids evolved from less developed species because of certain needs, and awareness of those needs. Quantum elements of the body (gnomes in particular) have been rearranged and modified by effort of the consciousness from the pool of infinite possibility. These choices are the result of an awareness of the need for change.

This awareness wasn't necessarily a conscious measure, but in most cases an unconscious safety change. These changes continue today. Our blood pressure, heart rate, respiration, temperature and so forth are all unconscious mechanisms that over periods of time could signal changes in individual genetics. Generally these changes would occur if there was some advantage in the modification.

Remember that the sensation of awareness goes all the way down to the cell and beyond. Chemical bonding is a very basic form of awareness. Certain elements prefer to complete their electron *shells*", and will gladly share them with other elements. This sharing forms a molecule of a new chemical.

It is my belief that body and soul develop together. They are dimensional twins that evolve with the universe. The physical body is changeable like a worn out suit. That thought may be upsetting to some, but the real *you* is the immortal one. The physical twin is immersed in space-time, and the other more closely allied with the infinite information of the non-physical dimension.

Injury or disease can disrupt the chemical-electrical functioning of the physical self, but the twin remains in its own element. Death is the ultimate malfunction of a physical entity. At this point, the quantum spirit is free of the labor of maintaining its extension in our material world.

Keep in mind that I am speaking now in the realm of metaphysics. These are subjects about physics, with a basis in fact, but as yet experimentally not proved. As I have mentioned before, because of my work with altered states the reincarnating spirit does make sense to me. Separation of body and spirit would bring with it a profound sense of freedom. The mind would no longer be encumbered with the restriction of material. Initially this freedom would probably be confusing.

Instantaneous transport and the availability of at least the past could be difficult for individuals to grasp right away.

Even more confusing would be the sense of being a part of everything. This sensation would be the direct result of finding that particles (quantum stuff) are holistically connected to each other. This is so even though the physical separation may be enormous. Recall that Russian experiment with the mother rabbit. Remember also that in the quantum (spiritual) realm, time and space have no meaning.

The great minds of our scientific world have stated that the past exists. Using the mathematics and laws of physics, this statement can be made and proven. For the general population however it does require some thought. If the past still exists, why can't we still see it, or feel it? We do live in a dimension of time; shouldn't it be available to us? We forget that many of the things of the past *are* sometimes available. These are the mountains and trees, the oceans and rivers, all of the things I have mentioned before that appears as more or less permanent fixtures. These things are constantly in the consciousness of a multitude of entities. All of the kingdoms of living matter are aware of these things, and thereby maintain the amplitude of its possibility wave. I have mentioned before that the stronger the quantum wave is, the more matter-like it becomes.

Many of the daily happenings of life are transitory. Things are not the same day after day. These changes of cognition alter the phase of the quantum waves and the interference patterns continue to reduce the amplitude of those waves of information.

A bird that flies by and lands on your fence post will never do it again in exactly the same way. The possibility is very small, so the material aspect of the bird is not maintained as it is for the post. The bird's flight isn't a permanent thing because it doesn't repeat in consciousness. It does still exist however as a minor wave or eddy in the complex chaos of the possibility dimension.

A memory is a transient reality. It is something that *was* and is reduced by a lack of repetition. It's much like the ring of a bell. If you strike the bell once it will ring with less and less amplitude until it is lost in time. A bell struck many times is noticed easily as a longer lasting thing, but once you stop, the sound is lost after a time.

I have mentioned before that altered states of consciousness increase our ability to recover those tiny waves of past reality. As you drift more and more into the non-physical world, the timelessness of that realm allows you to re-establish in your *mind* things that occurred and then moved on with time. I have emphasized mind, because these past occurrences are no longer in your brain, they are only available to your long term non-physical memory.

Those of you who have made use of meditative altered states can attest I'm sure to the clarity and vividness of visualization. Disassociation with physical awareness (relaxation) leads to greater and greater *reality* of your mental wanderings. Night dreams are a product of an unconscious state. To a dreamer, occurrences are real, because your ability to test for reality has been bypassed. Your body even reacts to that reality. Secretions ebb and flow in response to the stress and emotion of the dream. Respiration, heart rate and blood pressures change as the dream reality progresses.

Evolution has seen to it that the major muscles of the body are deactivated during these dreams or else we would walk around in response to the dream scene. Some of you right now are wondering about sleep walkers. Those individuals are not really asleep. Most sleep walkers and talkers come up from a deep Delta or stage four sleep into somnambulism, which is really a deep altered state of consciousness. In this state of hypnosis the protective muscle deactivation is bypassed.

In all of these states you are reacting to a different reality. What is the location of this reality? Individuals, who believe that the brain and mind are synonymous, believe the brain is generating its own little pallet to exercise the brain while the body rests. They can quite rightly point to all of the chemical and electrical activity that occurs during REM sleep. There is no doubt that all of these neural changes form our dreams, but from whence comes the information? Much of the dream information is new to you, so there is no reason to believe it is from memory. If you contend that your dreams are imagined information, it still must be information known to you but is rearranged. It is my belief that material memory and quantum possibility are both used to construct our dreams. Physiologists and neurologists are still working very hard to understand the nature of memory. Is it a function of chemistry? Chemistry eventually breaks down to an electrical function as we move down the ladder to the atoms and below. Are memories merely vast collections of potential variations? Does our brain function like a computer? Are we merely billions of on/off, plus/minus, 1/0, go/no go circuits? We know that isn't so because we can connect abstract things and principals. Computer specialists are trying right now though to make a computer that can do those things. Until we make a little more progress in our research, we must admit there is that possibility.

CONCLUSIONS

The purely physical memory looks less likely when we once again take note of the quantum functions of the electro-chemical basis of the brain. If electrons for instance were particles, this one dimensional brain mechanism would be possible. Electrons however aren't *things* they are inter-dimensional pathways that we can call gates, or pathways, or worm holes to a dimension of possibility and information.

If you recall your dreams, you will remember some of them as real theatrical productions. They have a plot and characters most of which you don't know. What is it or where is it in your brain that builds this story, and assigns the actors? Some of your dreams will have familiar places, things and individuals. If you really study some of those dreams though you may find that you recognize individuals not by physical features, but by traits and emotions.

I have tried to make it clear throughout this book that matter is a function of our consciousness. Our sense system is so arranged that concentrations of *information* are interpreted as *things*. That information is extracted from the quantum realm of possibility by acts of conscious choice. Most of these choices are made in combination with all other types of awareness. These are other people, animals, plants and so forth.

If we say that memories are stored in the brain chemically, we must also say that they are stored electrically. Chemicals are merely arrangements of electrons, protons and neutrons. All of these things are quantum waves or fields of information. This other dimensional information is infinite in scope, and available if our conscious effort properly phases the field.

It's a different concept I know, but stick with me. There is very little argument any more in the field of quantum mechanics, that consciousness effects our reality. Those experiments that are designed to investigate the velocity and position of an electron have shown this to be true. Also a particular thought process can extract information from the quantum field, and make it seem to be an electron particle. Different thoughts can make this same information into a wave. Part of our understanding must be that the wave and particle are the same thing. Both aspects of this information, whether we are talking about electrons or photons or any quantum particle, both particle and wave are *immaterial.* We sense merely the peak amplitude of in phase waves.

By following the last few pages carefully you can see that memory also is not a physical thing. The material of our brain, like the rest of our body is an invention of consciousness. Each atom of the material of our brain is constructed from quantum waves/particles. Experimentation has shown that these entities are not solid objects, but will manifest as one or the other depending on what we wish. In "The Matter Myth" two very, well known scientists, Paul Davies and John Gribbin discuss the solidity of matter when they say: "Quantum physics undermines materialism because it reveals that matter has far less *substance* than we might believe." The people that tell us these things aren't some group of radical spiritualists, but are very staid and conservative experimental physicists.

Since we can't really assign a place for memory to reside, we must assume that our experiences become a part of that field of information that we call the quantum dimension. Logically that makes it seem that all centers of consciousness are lines of experience into that infinite pool. That sounds very much like information reported by Paul Davies in his "Mind of God."

We have reached a point in our discussion of consciousness and the quantum that is inevitable. That is the question of personality survival of that state we call death. The question of survival really can be, deduced from the science of today. The foregoing pages have shown that the solidity of ourselves and our surroundings is sensory illusion. If we are not directly involved in experimental quantum physics, we can at least read of this progress. Careful analysis of these books and technical articles will indicate, as I have shown here that space-time activity isn't just four dimensional. Added dimensions (how many are still in question) will change our view of reality.

Our bodies consist of billions and billions of quantum points. These points are apparent crossover points to other dimensions. Since the experiments of quantum mechanics seem to indicate this, we are really multi-dimensional beings. We exist in space-time, but also extend into higher dimensions since our *bodies* are constructs of quantum waves. Our space-time reality is a function of our need. Conscious manipulation of matter is accomplished by the full spectrum of *awareness*. This awareness encompasses all of nature. This means from you and me all the way down to the atomic components that sense attractive and repulsive fields.

These multi-dimensional bodies that we like to think of as our own, really are only large groups of quanta working together at the direction of your individual consciousness. Why do they exist at all? It is my opinion that our bodies exist as a bridge from space-time to that dimension that is infinite information and knowledge. John Bell proved mathematically that the other dimensions are timeless. Because of the way we think, any place that is timeless is also space less. Every where and every when are a mathematical point.

While you have your mind in this scientific mode, think about life and death. A moment ago I said that quantum particles are a bridge between dimensions, and that we exist in both. It's sort of like a telephone line, except you are on both ends of the line. If you hang up on this physical end of the line, the other end is still functioning in the quantum dimensions. Consideration of the available information indicates that our so called physical body is a terminal point for information gathering in the segment of the universe that is concerned with space-time information. A defect that terminates transmission from our physical end of the line (death) will be a temporary interruption only. Your spiritual self (your entire being) will continue to function. As a resident of the quantum dimensions (spiritual side), you will have unlimited access to the past. As a non-physical being, you cannot effect the physical in any way, so there is no danger of altering the past. You would be able to re-examine your own past. I'm sure that will be true of all of your past physical cycles. In a timeless state you would merely have to concentrate and the information you wish to see would be available.

This quantum state I sense to be a dimension of intense curiosity. With infinite possibility available, I think a consciousness would ponder the same questions that we do, such as . . . 'What would happen if I did this?' This could be a superior consciousness, or just the total essence of universal intelligence.

As we live and experience, we constantly add to the universal intelligence. At the same time, our individual essence attains greater peace with that knowledge.

After death there is a time of rest and contemplation, and an examination of our input of experience and knowledge. Sooner or later curiosity would begin again and a search would start for a new physical side for your spiritual body.

Your search will end with a new body that will best be able to test your questions. Your experience will begin again with a new vehicle. Different quantum particles will have gathered together from the earth and sea to form this new body.

The memories of the previous bodies still reside in the universal intelligence of the quantum dimensions. These memories are only available if you temporarily put aside your physical presence. Learn to relax and enjoy all of your lifetimes. There is nothing to fear from life, or death.

PSYCHIC PHENOMENA

Some years ago Stewart Edward White published a series of books that told of his Psychic experiments in conjunction with his wife Betty, and a friend Joan. His most famous book was "The Unobstructed Universe." In his previous books, Betty in an altered state acted as a medium and offered information about the other side of the life and death duality. After her death, Betty communicated with Edward through their friend Joan.

The conditions described by Betty about this non-physical part of our existence, very closely resembles what we would expect in a consciousness controlled quantum world.

"There is only one universe," she said many times. "But there are many viewpoints of it." Life and death are two different aspects of the same thing. If you take a black and white picture of some scene and a color picture of the same scene, they are the same universe, but different aspects.

Betty would say, "The only difference between you and me is our awareness mechanism." This difference of *awareness* seemed to be ascribed to her sensitivity to exceedingly high frequencies.

I have mentioned that our material world is formed from very large amplitude waves of possibility. The large amplitudes are formed by infinite numbers of small waves coming *in phase* so they add. It seems that in the quantum (spiritual) world, all of the possibility waves are within their awareness if it is so desired.

Betty described our physical world as an "obstructed universe." Our physical sense mechanisms working through our consciousness increase the wave amplitude to a point of exclusion. We are aware of the very large amplitude waves (our matter), but the smaller waves of information are excluded from our awareness. The large waves of quantum field energy (once again our matter), compose our body.

The same high amplitude high energy waves compose our walls, trees, rocks and so forth. These waves of force are mutually exclusive following the laws of physics. This is why we bump into things. We are composed of maximized waves as are other material things, and our forces repel each other.

A wall is a great mass of electrons, within many combinations of atomic structures. To be in our universe they are maximum waves of quantum possibilities. If we try to walk through the wall we are stopped, or excluded because of that repulsive force that is exclusive to our space-time dimension. Our limited physical consciousness can't adjust our body to the higher frequencies of the out of phase waves.

We live in a dimension that is obstructed. Our shade trees obstruct direct sunlight. Roofs obstruct the rain, and anything physical is an obstruction to other physical things. As you move down the scale of atomic weight (fewer electrons, protons and neutrons) the obstruction gets less and less.

When our hopefully old body has stopped functioning because it has bumped into too many obstructions, all of a sudden we become aware of the "Unobstructed Universe."

The stories we read about near death experiences and that travel through the tunnel could be an awareness of the move towards that universe without obstructions.

Many of the so called psychic phenomena of our world can be explained by the laws of the quanta. Psychometry for instance is an outcome of the communication of electrons. Once a particle (any quanta) is in contact with another, they are in contact forever and at any distance. A sensitive individual can touch an item and make themselves a part of that chain. A psychometrist, knowingly or unknowingly resorts to an altered state of consciousness, availing himself (or herself) of information of the past or present. This information has become part of the quantum field.

The accuracy of predictions must be confined to areas of possibility. During our discussion on these pages, we have speculated that the future is formed by conscious activity. Consequently, other than mountains and planets the future is very plastic and is uncertain until the last instants. For this reason it is understandable that foretelling the future is always inaccurate by degrees. This would have to be true unless the future really is already in place for us. I for one dislike the thought of being an actor following a script.

In the same manner it is obvious that communication with surviving personalities in the quantum dimension is difficult. This type of communication is possible (presently) only with an altered state of consciousness. This altered state must be deep enough to eliminate the intrusion of conscious thought, and yet maintain certain physical abilities. Since control of consciousness is, subjective, true communication is still undependable.

The worlds of matter and spirit are one and the same. Growth of understanding, and the inevitable evolution of our ability to use our brain to its full extent, will make our world a pleasant place to live.

ONE UNIVERSE, TWO ASPECTS

The quantum particles of our body are the transceivers of the inter-connecting dimensions. The process of thought causes electron movement in our brain. Since the electrons are a part of the physical world, and at the same time are part of the quantum world of possibility, there is a change in both dimensions.

Our consciousness (thought process) has an effect on the dimension of infinite possibility and begins to oscillate possibility waves in phase with one another. Their strength and amplitude begin to increase. At some point the increasing possibility becomes a probability. Additional concentration of conscious focusing turns probability into powerful quantum particles that are sensed as matter. This matter is only one aspect of the continuum. All of this increasing strength of the possibility wave to make it matter-like is instantaneous.

It is essential in this phase of our evolution that we begin to understand our immortality.

Space and matter are both a function of time. In space-time (our universe) something that oscillates too rapidly to be sensed, or whose amplitude is too small, is not considered part of this world. Consider the virtual particles as a case in point. We call them particles, but don't consider their mass as part of this universe. We say that they emerge from nowhere and return to nowhere so rapidly, that their mass and energy do not effect our world, (not quite true).

Those ghosts like entities are merely a part of the quantum field that fills our entire universe. Since only our three dimensions are a function of time, and our sense systems are tuned to the amplitudes and frequencies of those aspects, we are generally unaware of other dimensions. We know only part of reality.

There are many degrees of consciousness (awareness). There is a consciousness of a cell, as there is of a blade of grass. Consciousness abilities grow as you climb the ladder of evolution. The awareness of an animal is less than that of a human. I speak now of a range of awareness. Animals are *more* aware of some aspects of our physical world. These sensitivities were evolved to enable them to survive in this world. Many times we hear people say that their pet sensed a tragedy before anyone else knew. This apparently is a function of their evolution. Animals are at a level that requires they pay attention to non-physical information. Their consciousness is more attuned to quantum information than is ours. Experimenters witnessed this when they monitored the mother rabbit in the Russian test I spoke of earlier.

Our lack of sensitivity isn't the lack of ability; we are just too absorbed with physical sensation. We become much more attuned to the other aspects of the universe during altered states of consciousness. If you recall I stated that our nervous system reduces physical sensation as we enter altered states. For this reason we begin to pay more attention to non-physical information that we receive.

The hypnotic inductions that I have used in my business take varying amounts of time. This is true mostly because individuals are sometimes more, or sometimes less ready to try something new. Some can go into deep states instantly. I have shown these phenomena on television several times. An individual need not appear to be hypnotized to be in an altered state. When you walk and day dream you are in an altered state of consciousness. An individual can drive many miles and be unaware of the distance traveled. It comes as a shock sometimes to realize you are at your destination and don't remember much of the trip. The lack of stimulation had altered your state. My students would remember instances like that and worry that during those times they were not in good control of the car. I assured them that isn't always the case. You enter the altered state because every movement is repetitious and has become a part of your unconscious, just like riding a bicycle. When anything unusual occurs, such as a car entering the road ahead of you, your state hopefully alters once again to fully conscious.

It is important that as we near a new millennium, that we begin to understand that consciousness is the only reality. We are aware of the four dimensional space-time, but we are only a little bit aware of the quantum unobstructed dimension. In our material world we are obstructed by material, by space and by time. These restrictions occur because of universal laws of physics. At some point we may understand the origin of these laws.

We are somewhat aware of the other aspects of our universe. During the past century we have been able to recognize and use electro-magnetic waves to our advantage. They have always been with us, but we walked right through them. We are in the habit of calling electricity "electrical potential." Quantum possibility is also a "potential." We have discussed that electromagnetic cycles are a 360 degree rotation of potential. When you have billions of electrons (or any quantum point) all rotating in phase, the potential can be great. The same is true of the *other* aspect of quantum particles. Throughout this book I have described the non-material aspect of particles as waves of *possibility*.

In our past history, conscious thought has visualized and made material the basics of our obstructed world. More time and deeper thought helped us construct machines and instruments from this basic material. The consciousness of individuals influenced the phasing of waves of possibility to increase their amplitude and make them material machines. Not just by thinking about it, but by using material that has been made *solid* by other thought processes.

In one way it is a very simple process, in another a very convoluted and technical process. From mental images come material things. The images have been borrowed from the quantum realm of infinite possibility.

Generators and dynamos produce usable energy by manipulating magnetic fields and electrons. Few people stop to think that the idea *itself* was the manipulation of electrons in the brain influencing the quantum field.

I'm sure that it is going to take some time for our people of science to recognize that life is multi-dimensional. Our universe is much more than we realize. There is a component of the universe that is obstructed, but all of these obstructions are self imposed. They are a function of our combined consciousness. Our physical brain is a biological mechanism that operates as a receiver of information. Sensation arrives from our physical domain, and information can be obtained from both our physical and non-physical dimensions.

Physical sensation is part of our obstructed world and cognition and information can be a part of both worlds. Each time a physical entity has a sensation of any kind, the electrical functions in the brain are recognized at the quantum level. This recognition effects the phasing of quantum waves, and emphasizes the sensation.

Remember that sensation is consciousness, even if the sensation is non-physical such as dreams or hallucinations. Experimental physicists have shown that consciousness is a quantum effect.

Our various states of consciousness can be and will be used in the future to study the entire continuum of physical-spiritual existence.

It is necessary that we begin to recognize that our body and consciousness are an integral part of the universe. We exist in a wide range of dimensions. Dissolution of our physical body is important only to our physical family. In time, we will learn that it is really a necessary change, and that we should be glad of the time we shared in this material world. The quantum particles continue, and at that point consciousness will no longer be obstructed by matter (space-time). We will not be restricted to this world if we wish new sensations of a physical nature on other worlds. The universe is open to our choice.

The quantum researchers are beginning to accept the theory that all entities are universally one. They reason that since Bell's theorem shows that there is instantaneous communication from one end of the universe to the other, there is but one unit. High powered mathematics reveals that at some time between the Big Bang and now that all quantum particles have been in intimate contact. Because of this contact, Quantum laws indicate that all particles are aware of the condition of each other. Quantum activity on one end of the universe is instantly recognized at the other. Keep in mind that time is not an obstruction in the quantum (spiritual) dimension. Remember our little discussion about how the two coasts of the U.S.A. seemed to get closer together as the speed of transport increased? When everything seems to be in the same spot, communication is easy. If an electron can be identified as a mathematical point, within that point are all of time and space. Matter and time are all represented as possibilities.

Our material obstructed world is one aspect of the continuum. The quantum un-obstructed world of possibility is another aspect. When we begin to contemplate ourselves, our friends and family, we must begin to recognize that our *presence* isn't merely physical. We have an aspect of solidity in this material world. Existing with that physical presence is another that we call consciousness. They are the same presence, but in the material world the quantum particles appear as an obstruction. The obstruction is a consequence of low frequency (because of mass) and large amplitude waves resulting from phase relationships between centers of consciousness.

When circumstance makes it difficult for a consciousness to maintain its outer shell in the physical world, it ceases to concentrate on that maintenance. The physical essence of that consciousness will gradually decompose from that point. The dense material such as bone will change more slowly for various chemical reasons, but these are still quantum changes. The loss of this physical shell has nothing to do with the immortal consciousness. Remember that a physical eternity means nothing in a timeless realm. If you read the books and articles written by psychic and sensitive people, you will notice that there seems to be time and material involved in the spiritual world. This could indicate that a consciousness that is living only in the realm of possibility will create a psychological time by manipulating those possibilities.

There is an extreme amount of scientific research that indicates that our bodies, as well as all material things are made of quantum particles derived from waves. In the case of our bodies, they are maintained by a concentration of consciousness. Individual particles, atoms, elements and cells all cooperate to afford the consciousness an extension of itself into the world of space-time. When material not within the control of that consciousness invades the body, often the space-time extension is destroyed. This *death* of a body is similar, and not much more important than the breakdown and loss of a familiar car. The controlling consciousness still exists in the quantum world of universal possibility.

I'm often chastised by my immediate family for my lack of enthusiasm for religion of any sort. I try to stress that I do, it just isn't structured like theirs is.

My religion is universe wide. Each and every point of consciousness is a part of an expanding intelligence. Contemplation of the magnificence of nature and that each of us has a part in its physical maintenance continues to drive me. All of the other conscious entities of the universe in their own way also strive to increase the total intelligence of the universal consciousness by their choices. Each action that effects the phasing of the quantum possibilities to make it manifest in the space-time world, moves the two attributes of the universe closer together. Greater understanding will promote greater communication between dimensions.

Main stream physicists that are employed by our large universities or big business either say little, or object weakly to metaphysical concepts that speculate on the continuity of consciousness. This is so even though most will admit that the universe seems to be a total consciousness. Physicists' Paul Davies and John Gribbin make a good case for this in their book "The Matter Myth." The reticence of some can be understood, since the grants and university seats depend on sticking to the current political right.

SPECULATIONS

The need for grants and the laboratory of universities has silenced much of the enthusiasm for Bell and Aspects findings that reality as we know it does not exist, at least at the sub-atomic level. If the quantum findings are extended to our macro world, some cherished beliefs must be abandoned.

In 1982, Alain Aspect, Jean Dalibard and Gerard Roger proved through experiment, that either objective reality does not exist beyond the mind of the observer, or faster than light communication with future and past is possible.

These experiments were carried out by using two photons of light emitted simultaneously by an atom. See "God And The New Physics", by Paul Davies. Aspect maintains that one of the two alternatives must be fact. I prefer to believe that both are true, modified only by degree. These experiments, although truly historic, were met by a strange silence. This lack of enthusiasm is for the reason I gave a moment ago. Main line scientists fear being labeled *far out* or *fringe* when grants are considered.

We have discussed much about the electron, but have neglected the "spin" of an electron. Sub atomic particles have been assigned different spins to account for experimental results. The electron has been assigned a spin of 1/2. Unlike a sphere in our macro world that makes one complete rotation in 360 degrees, an electron requires 720 degrees for one complete rotation. This is difficult to visualize unless you speculate as I do. Throughout these pages I have said that our physical space-time world is intimately connected to another dimension that we have been calling the quantum dimension. This other dimension is free of the restraints of space and time. This freedom allows two aspects of material, and non-material to exist in what seems to be the same space. I think that this so not only for the very small like the electron, but also for ourselves as body and soul.

I believe that the electro-magnetic force of the electron makes a 720 degree rotation around the two phases of the electron, in much the same fashion as our sign for infinity, the lazy eight.

It has been found that the magnetic field produced by an electron is exactly twice what would be expected from a spinning charged sphere.

In quantum physics, it is meaningless to think of a particle having a trajectory through space. It is also meaningless to think of the force of the electron rotating 720 degrees without half of the rotation being in a different dimension.

The experiments that have been conducted to ascertain whether electrons are particles or waves have shown that they can be both. I speculate that during transit from electron gun to sensor (in the lab), that the electron is in non-space, and can adjust itself to one path or to multiple paths. This would be a possible explanation for the interference pattern in the wave/particle experiments.

Throughout these pages I have described the results of many experiments conducted by the world's imminent physicists. These experiments that you can read for yourself (see bibliography) have inevitably shown that the matter that we are so fond of is quite ethereal itself. We have seen that quantum particles have dual existence, and that they exist as mathematical points in our space-time universe. Strangely, a mathematical point is supposed to occupy no space at all. This makes me believe that each quantum is a focal point of the possibility field from outside of space-time.

Many eminent physicists believe that this field is consciousness, and that our universe is composed of one great consciousness. Menas Kafatos and Robert Nadeau say in "The Conscious Universe", "The universe on the most fundamental level is an un-dissectible whole."

Elsewhere in these pages I described quantum points as wormholes, or tunnels from space-time to other dimensions. Since all matter is composed of quantum particles (points) it follows that all things, including you and me, exist in more than four dimensions.

Bells' theorem has proved that at one time all particles were in intimate contact and have remained in instantaneous contact since.

A thought process in our physical brain involves the movement of billions of electrons. As these electrons move in intricate patterns, their electromagnetic force disturbs space-time as well as their counterparts in the quantum dimension of possibility. Our experiences in this universe are also imprinted in the universal consciousness.

This is that place of the quantum dimension that we call information, or possibility. This universal information (your memories and mine), is available to everyone that can develop the talent to extract it. Examples of this are seen everyday. Individuals that practice psychometry are gaining small bits of information about other individuals, by handling items that have been in intimate contact with the other person. Although the practice of psychometry is still not viewed as a verifiable talent, many police departments throughout the country have used these sensitive's to aid their investigations. As I have been saying, all quantum particles seem to act as a component of a universal mind or consciousness. Consequently memories and sensations are available to all as part of their own.

A method of connection or transmission to the other dimension of our being is through the virtual particles. As we receive information, there is a frenzied movement of electrons in our brain. When we react to that information, there is also a huge movement of electrons. This causes a displacement movement in the virtual particles in adjacent space and non-space.

Each electron is accompanied by an electromagnetic field that tends to push away like charges. By virtue of this transmitted force, a similar pattern is established in the dimension opposite our space-time universe. It is the dimension of consciousness.

Part of your path to being comfortable with life and death is to understand that we *are* immortal. Our immortality isn't ritualistic and dependent on ancient codes and mores. Kafatos and Nadeau state that "the world view of modern physics is more consistent with eastern metaphysics." Although well known physicists like David Bohm and Fritjof Capra have drawn these parallels, many do not think that the comparison is valid yet.

If you become aware of the research and findings of our physicists, you will begin to feel that the quantum dimensions are the basis and consciousness of the universe. Our existence now and later will be comfortable and pleasant if we stay aware of the rest of our universal self.

I believe that our individual consciousness emerges from time to time in this material world to learn and react to various situations that have no counterpart in the unobstructed world. In this way the universal consciousness gains information that will be stored and made available to all. Because of the nature of the unobstructed universe, I also believe that useful information is available from all of the worlds of the universe.

Cosmologists such as the late Carl Sagan can calculate the possibility of billions of habitable planets in the universe. It would be extremely egotistical of us to believe we hold any special place in the order of things. Many would argue that the same point would be true even if we believe that we are immortal. There is however a difference in a belief of *faith* and one based on experimental science.

Earlier we discussed briefly that our bodies are composed of ancient materials. These materials were absorbed from the water and soil in a way that was prescribed by our DNA codes. Once again this is quantum material that is arranged in code form. The code is meant to supply that particular body with talents and shortcomings that will enrich the consciousness with new information and sensation. All of this becomes part of the universal organism that is multi-dimensional.

The title of this chapter is "Speculations" because I wanted to relate verifiable information that I can use to speculate on conditions of reality.

The scientifically verifiable information comes from the scientific journals of the day, such as those I have listed in the bibliography. The speculation is mine alone.

For instance we are aware that *quanta* really have no definite location. They seem to be indefinite and malleable. The result of the interaction of their waves or fields gives us what we consider to be *matter*. We are aware of this matter because of our consciousness. Our body is a quantum field of maximized waves. If, as some profess consciousness arises only from the workings of the brain, consciousness is indeed a non-material field. I believe that the brain is a mechanism, highly adapted to extract information from the information realm of the quanta. At the same time this mechanism will insert sensations into the same quantum field. Our memories aren't stored electrically in our brain, but flow through from elsewhere, controlled by the frequency and wave shape modulating system of our brain.

Few people really stop to think about some of their sense impressions. A walk in the garden can be a good illustration. We can be mildly aware of the effort of walking, and the sound of our footfalls on a gravel path is also a background sensation. Our main center of attention however would be sight sensations.

The complex pattern that we label a *rose* is transmitted to us by photons. These are information quanta that our eyes and brain transform into very large numbers of frequency and amplitude variations. This gives us color, size and shape. The continuous stream of photons relates to us changes that occur while we are conscious of the rose. This gives us a time sequence, to be aware of leaves and petals moving.

These information signals are all non-material bits of quantum information. This information is received by your eyes and brain, and is compared to similar patterns that our particular consciousness has received before. This process is entirely non material. I have to repeat once again that physicists tell us that, the electron is not a thing. The rose and our sense equipment including our brain are composed entirely of electrons, protons and neutrons. All of these are non-things, and are really points of quantum possibility. The photons that carry the information are also quantum particles.

The consciousness that becomes aware of these colors, shapes and time changes is also non-material. At any later time you could destroy the sense receptors, the brain and the body, but that information of the space-time dimension is retained by the consciousness. This information is eternal in the quantum dimension of information since there is no time division in that dimension.

Often when I would lecture, people would question the concept of a universal consciousness. Their main point was a kind of worry that their individual egos and thoughts would be lost at the change we call death.

Once again I would have to speculate and say that it *seems* that individual consciousness remains in a state that can be compared to the construction of the human body. Each cell is an individual thing, but is a part of a larger entity. The same could be true of individual egos. Each is a part of "The Mind Of God", a speculation by Paul Davies. A somewhat less spiritual way to put it would be to consider a consciousness unhindered by a physical body. This entity could be concerned with self and its memories. At the same time it would be a part of the super conscious. The super conscious would be an entity that is the sum total of all experiences and intelligence available from all sources of consciousness.

It is good to remember that these sources of intelligence are universe wide. This is explained in "The Conscious Universe", by Menas Kafatos and Robert Nadeau. The quantum field from which our possibilities are developed is not constrained by time or space.

You will recall that there are possibly billions of habitable planets available for the development of sentient beings that are the physical part of a consciousness. Each one of these can contribute to the total consciousness. Whenever someone has a great idea, or a new concept, it comes from a place where all possibilities exist.

Spirit and Body

Physical reality as disclosed by quantum physics is non-local. As we have speculated, non-local really means universe wide. It is a fact of nature and it shocks most physicists. It takes away permanently the bias that our physical world is composed of individual objects. We must amend our thinking to allow *all* objects to be part of the whole.

John Bell's theorem and the Aspect experiments have forced us to realize the modern physics overview resembles eastern metaphysics. Buddhism, Taoism and Hinduism; these all see the universe as one evolving unit, but each has its own view of the process.

The duality of existence has been with us from the beginning. Consciousness is the driving force of evolution in all material things. There is some evidence that the Neanderthal and Cro-Magnon branches of the human tree each had some regard for burial. Objects have been found with some remains that indicate the survivors were concerned with the afterlife of the dead.

All of the religions of today are concerned with various ideas of survival. It has been customary for religions to equate the surviving entity with a non-physical state. One of the problems of religion is that the reality of this state must be accepted on faith alone. To complicate matters further, this spiritual state is reserved for only a select few. In many of them men, women and children who have not been anointed by the guardians of the faith are relegated to total annihilation, or are punished throughout eternity.

This sheep like following of a *shepherd* was probably fine for a population that was uneducated but is lacking in reality for educated masses. The stories of the various sects can now be replaced by repeatable tests that will remove the scientific view from the *theory* faze to one of fact. The writers of the scriptures of all religions used the knowledge that was available to them at that time. Experimental physicists are using the knowledge of today to show proof of a dual existence that is the true state of us all.

There is a point that I wish to mention that may need some future thought, and that is that we are not necessarily *the observers*, but possibly the observed. Remember that there are several layers of consciousness.

Our previous examinations have shown that a conscious observer will modify the results of an experiment of quantum materials. Maybe we are being a bit egotistical to believe that it is our small consciousness that is in control. In my studies I have found no reason why this control of experimental answers must originate in our space-time world. Werner Heisenberg believed that observation creates reality. At the same time there is no mention of the state of the observers. In the past we have all assumed that our thoughts manipulate things so that we can observe only one reality, such as locality at the expense of velocity. Why must that pressure to modify quantum waves come from our dimension?

I have speculated that our consciousness is but a tiny part of a consciousness that is the sum total of all entities. The laws that seem to limit our understanding of the true structure of the universe may be limits placed by a greater intelligence until we attain an ability to understand.

Maybe that unused 90% of our brain has to become activated. There could be a kind of distinction between the mechanism, and the operator. For a long time there has been discussion and controversy about the brain-mind duality. Some physically biased researchers insisting that all consciousness is a product of the brain. Other researchers believe that the brain is a mechanism that is a physical counterpart of a spiritual other dimensional mind.

During lectures in the past I have often compared this duality to our new adventures with space flight. For flights to other worlds' we build a mechanism and equip it with sensors. These sensors are chosen to give us specific information about this new place that we are investigating. The transport from home to this new place is controlled by the builders of the instrument. Sometimes advantage is taken of forces and fields that are in existence in route to this new place. Our rocket scientists have done this many times to give the satellites an additional boost of gravitation to increase the velocity of the satellite. After arrival, the mechanism tests the air and soil as well as the gravity and other forces. It sends this information back to home base.

There are circuits built into the mechanism to recognize some dangers and to avoid them. There are also some things that the machine is not equipped to understand or to react to. This situation is handled by the home base. Signals are sent to alter course, or to change its program.

The situation can be very similar to the brain mind duality. Signals are going in both directions. Some situations are handled by the mechanism (the satellite) and in our case (the brain). Some are handled by the home base (mind). Both are a part of the unit, but one is an expendable extension and the other a permanent area of understanding.

This analogy may help you to understand that our body is a mechanical extension of our consciousness into an alien world. This extension is endowed with a limited amount of ability on its own, but its limitations can be overcome with the proper communications with its builder. To carry the analogy further, the mind (our ego) is an extension of a still greater home base, which would be the next level of complexity and maturity.

As physical beings we employ instruments to investigate areas of space-time that would be otherwise unavailable to us. We use electron microscopes to see the very small, and various types of telescopes to investigate and contemplate the very large. It would be inevitable that a purely spiritual being would develop instruments to investigate other dimensions. We are those instruments and mechanisms, but are an extension of the spiritual body (consciousness) of the builder.

In this compound arrangement... man or woman to consciousness, consciousness to super conscious and super conscious to another step up; some physical happenings will come as a surprise to all levels. The character of free will insures that the future is the result of choices from the possibilities of the quantum dimension.

The uncertainty principal seems to be the law of the universe. To a large number of individuals the universe is a reflection of God, or the supreme consciousness. This supreme intelligence sits not in judgment, but in curiosity at the choices made. All of this being so, the future is uncertain. This concept of an uncertain future is known as deism. It allows a supreme being to begin a universe, but to have no control over its fate. *Theism* on the other hand is a belief that a supreme being creates the universe and continues to intervene in its progress.

In both of these belief systems, God and the universe are separate. The Supreme Being stands outside of the universe, and is something different.

In a belief known as *pantheism*, God and the universe are one. The supreme consciousness is everything. The nature of space-time and of the quantum dimensions are all part of the "Being of God." In this belief system, we are a part of the Supreme Being and are evolving with the universe.

The space-time segment or dimension of the universe is responsible for physical evolution as well as spiritual evolution. The arrow of time, dictated by the second law of thermodynamics, requires our spiritual selves to examine choices made in the past. These choices are not reversible, but become memories and wisdom we can rely on in times of future choices. In our physical life these memories aid us in the form of instinct and intuition.

We have discussed before that most of us prefer a world and universe of free will. In order to have that, we must live in a universe in which the designer *has* no control or has voluntarily given up control of the future.

This particular viewpoint seems to agree quite well with what we observe in Heisenberg's *uncertainty principal*. If all beings and material are a part of the supreme intelligence (Pantheism), and we are all evolving together, the future of the Earth and ourselves is in our control by the choices we make. Many of them that we are making are very bad, such as our over population and over use of greenhouse gasses. I fear for our survival. We are miss-using the Earth and it is becoming quite ill because of it. That topic however will require another book.

In the quantum world all things are possible. It is the same in this space-time world until choices are made. The vast majority of these choices are automatic. They are made necessary by the past. In some cases however the future is not decided until that instantaneous decision. The instantaneous communication across the universe posited by John Bell and Einstein's light speed limit on information across the universe poses a great contradiction. These contradictions make it necessary for extra dimensions, or at least a concept that everything exists at the same point.

Instantaneous means *no time*, and no time means *no space*. Without these two you have a mathematical point. It's like the old question, "How many angels will fit on the head of a pin?"

It seems that if you follow the rules of physics, everything is an illusion. Einstein spent his later years trying very hard to find a way out of the illusion concept.

If you are confused at this point, don't feel bad because the physicists are also. To help bring things together a little bit, here are some things to remember.

1. As of right now space is a "thing." it is a tangible part of our dimensions. Length, width and depth encompass "space."
2. Cosmologists such as George Gamow theorize that our universe originated from a point labeled a singularity.
3. A singularity contains all four of our dimensions compressed to zero size.
4. This mathematical point began to exhibit volume and time at a point called the *Big Bang*.
5. After a period of inflation (vastly accelerated growth because there was no material yet that required a standard velocity), space-time expanded at the speed of light.
6. The usual explanation for the eruption and expansion of space-time is that the point existed in an unstable vacuum.
7. It *is possible* to visualize our universe expanding *into* something if you recognize space-time as a unit, becoming a part of other dimensions.

There are many ways to contemplate the beginning of our universe. I prefer to think of it as the beginning of awareness for a new entity. The brand new universe was initially aware of heat, light and pressures. As the expansion slowed to a standard light speed, awareness grew to include large waves of possibility that were represented as particles.

Because these particles were once intimately joined in that unstable point (that was of other dimensions), they found each other attractive, and maintained a communication with all. These atoms of Hydrogen and Helium (mostly), began to gravitate together. As the particle/wave groups increased in size, their phase relationship became great enough to constitute a material *thing*.

The consciousness that is our universe continued to receive information about itself from its constituent parts. These parts were the initial atoms that were formed of quantum waves of electrons, protons and neutrons. It received information also from other dimensions through the quantum contact. You can ascribe as much intelligence to our universe as you like. It can be very basic, or it can be super intelligent. Remember however that the choices that the universe and you make are a phase relationship in the quantum dimension that becomes powerful enough to become matter-like.

The photon is considered a messenger particle, and carries information from one particle to another. The photon is also the measuring medium for time. These two things being true, doesn't it seem logical that the increase in volume of the universe can be a measure of increased awareness (consciousness)?

SUMMATION

The main thrust of this small book is to acquaint readers with some very basic outlines of consciousness and quantum physics, which are intimately connected.

These two studies actually require a great deal of expertise if they are to be your life work. They are however within the grasp of most everyone if they can be content with an overview, or partial explanation.

Consciousness goes by many names; I prefer awareness, because you must be aware to be conscious.

The question has always been, is that mind or spirit the result of the body, or is the body the result of the spirit? Does the death and dissolution of the body constitute the end of the mind or spirit? What happens to me when the body dies?

Physiologists, or at least those people that believe only in the *physical* things find it difficult if not impossible to believe that anything can exist without material properties. The life work of the various religious organizations is to convince people that physical things, our body included, are temporary and discardable.

We have these two extremes. One says the material body is all there is, and the other that the spirit *only* is important.

For many years the scientific community was leaning pretty much toward the physical side of the controversy. Even though many of the greats of science were skeptical of non-physical existence, they kept their religion as a backup. The religions of course insist that you accept their ideas *on faith*. The scientific community on the other hand insists on repeatable proof of any and all suppositions. The proof of a spirit or soul still eludes both religion and science, but at least science is looking.

There has been little progress over thousands of years towards actual *proof* of spirit, or consciousness. Even a *search* for this proof has been hampered by the various clergies. Essentially the reason for this is their feeling that it is sinful to question what is written.

In the early part of this century science greats' such as Einstein, Schrodenger, Borne and Heisenberg formulated what came to be known as "wave mechanics." Their experiments showed that certain aspects of matter always seem to be hidden from us. This fuzziness about some attributes of quantum particles turned out to be a definite law of quantum physics. The rule is, "If a choice is made, all other aspects become only possibilities."

If you think about it and transpose the rule to our macrocosm, it still holds true. Once a decision is made, and action taken, it becomes part of the past and is unchangeable. The other things you could have done remain only possibilities.

This rule is the result of other strangeness of the quantum realm. The so called *particles* of our universe seem to exist in a condition *between* states. Unless there is a consciousness that is making choices, or asking questions, the quantum entity is an ambiguous presence. It is neither a particle nor a wave. It is either a combination of both, or it oscillates between the two states so rapidly that it escapes our notice. The force of consciousness seems to have the ability to form the quantum entities into what we sense as material things.

Yes we do bump up against material things. The reason for this can be compared to magnetic poles being exclusive. They can either attract or repel, and it depends on the field.

Your hand can rest on the table top because of the exclusive *fields* of the table and your body. In reality, both are an *effect* of consciousness or spirit.

A quantum entity is a crossover point between space-time and those other dimensions that in our history we have considered *spiritual*. A quantum particle (electron, neutron, proton and photon) that is not being maximized by a consciousness is a neutral entity. It is both here and there but is inactive in both the material and non-material universes.

In earlier pages I have talked about the possibility waves. In the non-material areas of our universe, this is a field of infinite information. All things that are possible are available there. It is necessary however to amplify them to make them a reality in our material dimension.

This kind of mechanism can be compared to a range of radio frequencies. Our equipment, both radio and senses will pick out one range of waves and amplify them to reality. The rest of the infinite possibilities are still there simply awaiting the proper sensing equipment.

It is sometimes useful to me to visualize this procedure of choice from the quantum spiritual dimensions in a mechanical kind of way.

A thought process begins in our brain with the movement of billions of electrons shifting positions in the brains physical field. These electrons leave an impression in space-time, but also in the quantum realms via their 720 degrees of rotation. The forces of our material thoughts in this way enter into the field of infinite possibility and amplify one possibility to reality.

This arrangement seems to rule out a predetermined future. If the future were predetermined, there would be no need for a space-time dimension.

The intimate connection between our thoughts (consciousness) and the realities of our material world are obvious. Without an exchange of forces such as material to spiritual and spiritual to material, the universe as we know it would not exist.

Physicists such as Paul Davies and John Gribbin feel that the obvious effect consciousness has on the experiments with quantum particles, does not carry over into the macro world.

Why is it that these physicists and some cosmologists have no difficulty accepting the theory (big bang) that our entire physical universe emerged from a mathematical point and yet are unsure if the same quantum activity (on a much smaller scale) can effect us in our human scale?

The majority of the scientists that are experimenting to find the true status of matter agree that each bit of matter is a quantum entity. They also agree that an observing consciousness effects that material.

If an entire universe can emerge from a quantum point, why does it take such a leap of faith to suppose that consciousness emerged in that same instant?

If a tiny quantum area (an electron) is a field that we can detect (electro-magnetically) and is affected by an experimenter's thoughts, why should a compound area be any different? The mathematical qualifiers of wave length and mass are a factor only if a body (yours and mine) is considered as a unit. The human spirit isn't a unit of mass; it is a field of consciousness.

As your life unfolds remember that you have a great deal of control of that life by virtue of choice. It is true that other entities exert some influence by their choices, but *you* are still the captain of your soul.

Recognize that at this moment you inhabit both the material world and the quantum (spiritual) world. Your material self is temporary, but your other dimensional self is immortal. Actions taken here are reflected there. An eternity of remembering wrong choices should be an incentive to make as many right ones you can. Your spirit is a part of every material and wave in the universe. Help to maintain the health of our present home (the Earth) by recognizing that it is at the limit of its ability to support our population.

BIBLIOGRAPHY

The books mentioned here were helpful to me to gain a lay education in the rudiments of quantum theory and quantum mechanics. These books for the most part have a minimum of mathematics and are written for the general public.

Most of us will have to leave the technical work to those very talented physicists. Included are some books that have given me more insight into the very speculative psychic phenomena that is possibly explained by the findings of quantum science.

Brackman, John "speculations—The Reality Club"
    Edited by John Brackman
Capra, Fritjof "The Tao Of Physics"
Cassidy, David C. "Heisenberg, Uncertainty and the Quantum
    Revolution" Scientific American—May 1992
Davies, Paul "The Mind of God"
Davies, Paul "God & The New Physics
Davies, Paul and John Gribben "The Matter Myth"
DeBroglie, Louis "Matter and Light: The New Physics"
Dyson, Freeman "Infinite In All Directions"
Feynman, Richard P. "QED"
Gribben, John "The Omega Point"
Gribben, John "In search of Schrodingers Cat"

Gribben, John and Martin Rees "Cosmic Coincidences"
Herbert, Nick "Quantum Reality"
Johnson, Raynor C. "The Imprisoned Splendor"
Kaftos, Menas and Robert Nadeau "The Conscious Universe"
Leslie, John, editor of "Physical Cosmology and Philosophy"
McCreery, Charles "Psychical Phenomena And The Physical World"
Menas Kafatos & Robert Nadeau "The Conscious Universe"
Montagu, Ashley editor "Science and Creationism"
Ostrander, Sheila and Lynn Schroeder "Psychic Discoveries Behind the Iron Curtin"
Ross, Hugh "The Fingerprint Of God"
Segal, Julius and Gay Gaer Luce "Sleep, The Third World Of The Mind"
Talbot, Michael "Beyond The Quantum"
White, Stewart Edward "The Unobstructed Universe"
Zukav, Gary "The Dancing Wu Li Masters"

Printed in the United States
68800LVS00002B/31